THE THEORETICAL PRACTICES OF PHYSICS

The Theoretical Practices
of Physics

Philosophical Essays

R. I. G. HUGHES

OXFORD

UNIVERSITY PRESS

PHY S

OXFORD
UNIVERSITY PRESS

Great Clarendon Street, Oxford OX2 6DP

Oxford University Press is a department of the University of Oxford.
It furthers the University's objective of excellence in research, scholarship,
and education by publishing worldwide in

Oxford New York

Auckland Cape Town Dar es Salaam Hong Kong Karachi
Kuala Lumpur Madrid Melbourne Mexico City Nairobi
New Delhi Shanghai Taipei Toronto

With offices in

Argentina Austria Brazil Chile Czech Republic France Greece
Guatemala Hungary Italy Japan Poland Portugal Singapore
South Korea Switzerland Thailand Turkey Ukraine Vietnam

Oxford is a registered trade mark of Oxford University Press
in the UK and in certain other countries

Published in the United States
by Oxford University Press Inc., New York

British Library Cataloguing in Publication Data

Data available

Library of Congress Cataloging in Publication Data
Hughes, R. I. G.
The theoretical practices of physics : philosophical essays / R. I. G. Hughes.
p. cm.
Includes bibliographical references and index.
ISBN 978-0-19-954610-7 (hardback : alk. paper) 1. Physics—Philosophy. I. Title.
QC6.2.H84 2009
530.01—dc22   2009029698

Typeset by Laserwords Private Limited, Chennai, India
Printed in Great Britain
on acid-free paper by the
MPG Books Group, Bodmin and King's Lynn

ISBN 978-0-19-954610-7

1 3 5 7 9 10 8 6 4 2

For Barbara

Our mutual love of theatre brought us together one Spring night in the Autumn of our days, and together we have lived our own drama filled with joy, laughter, and love. Thank you, my Dear, for bringing a new, undreamed-of dimension to my life, and for your unfailing love, encouragement, and support. In the words of the Bard:

> "That time of year thou mayst in me behold
>
> When yellow leaves, or none, or few, do hang
>
> Upon those boughs which shake against the cold,
>
> Bare ruined choirs where late the sweet birds sang.
>
> In me thou seest the twilight of such day
>
> As after sunset fadeth in the west,
>
> Which by and by black night doth take away,
>
> Death's second self, that seals up all the rest.
>
> In me thou seest the glowing of such fire
>
> That on the ashes of his youth doth lie
>
> As the death-bed whereon it must expire,
>
> Consumed with that which it was nourished by.
>
> This thou perceiv'st, which makes thy love more strong,
>
> To love that well which thou must leave ere long."

William Shakespeare, Sonnet 73

Foreword

Over the last ten to twenty years, I have become increasingly interested in exploring the ways in which scientists, and physicists in particular, explain and communicate their theories to each other and to the world at large—in other words, the methods they practise in the discourse of physics. As a philosopher of physics, my interest and research in this area led me to the conclusion that I wanted to write a series of essays dealing with various facets of this subject, both philosophical and scientific, and then to tie them together in a book. This book is the result of this quest, and covers a wide range of subjects, from the application of literary criticism to the discussions and analysis of writings in classical and quantum mechanics.

The book consists of eight essays, all within the Philosophy of Physics, and varying in length. They fall into three categories. In the first category are the essays which examine specific branches of physics: Essay 2 covers Bohm and Pines' papers on the plasma theory of metals in the early 1950s; Essay 6, Kenneth Wilson's critical-point phenomena up to 2000; and Essay 8, the Aharonov-Bohm effect. In the second category are the essays that concentrate on facets of physics that philosophers attend to: laws of physics in Essay 3; disunities in physics in Essay 4; models and representation in Essay 5; and theoretical explanations in Essay 7. In the third category is Essay 1, a hybrid essay built from criticism and the general theory of relativity.

Essay 1 introduces criticism, not as 'disapproval', but as 'the definition and judgement of literary or artistic work'. The literature discussed is Newton's *Principia* (Essay 3); Newton's *Opticks* (Essay 7); Einstein's papers on *The Brownian Motion* (Essay 4); and Einstein's *The Perihelion Motion of Mercury by the General Theory of Relativity* (Essay 1). Essay 2 introduces 'theoretical practice'. It provides a commentary on Bohm and Pines' quartet, *A Collective Description of Electron Interaction*. In the quartet, Bohm and Pines employ (a) two theories: classical mechanics and quantum mechanics; (b) sundry models; (c) several modes of description; and (d) various approximative strategies. By treating these publications as texts, it casts the philosopher of science in the role of literary critic.

When tied together, the eight essays form the whole, a discussion of the *theoretical practices of physics*.

Contents

List of Figures and Tables

Figures

Tables

List of Abbreviations

AT	René Descartes, *Œuvres de Descartes*, ed. Charles Adam and Paul Tannery, 11 vols. (Paris: Librairie Philosophique J. Vrin, 1983), vol. vi
BCS	Bardeen, Cooper, and Schrieffer
BP	Bohm–Pines
BP I	David Bohm and David Pines, 'A Collective Description of Electron Interactions. I: Magnetic Interactions', *Physical Review* (1951), 82: 625–34
BP II	David Bohm and David Pines, 'A Collective Description of Electron Interactions. II: Collective *vs* Individual Particle Aspects of the Interactions', *Physical Review* (1952), 85: 338–53
BP III	David Bohm and David Pines, 'A Collective Description of Electron Interactions. III: Coulomb Interactions in a Degenerate Electron Gas', *Physical Review* (1953), 92: 609–25
BP IV	'A Collective Description of Electron Interactions. IV: Electron Interactions in Metals', *Physical Review* (1953), 92: 626–36
CA	cellular automaton
DDI	'denote', 'demonstrate', and 'interpret'
GTR	general theory of relativity
P IV	David Pines, 'A Collective Description of Electron Interactions. IV: Electron Interactions in Metals', *Physical Review* (1953), 92: 626–36
P	Isaac Newton, *Philosophiae Naturalis Principia Mathematica*, ed. F. Cajori (3rd edn, Berkeley, Calif.: University of California Press, 1934)
PSSC	Physical Sciences Study Committee
PT	Pierre Duhem, *The Aim and Structure of Physical Theory*, trans. P. P. Wiener. (2nd edn, Princeton, NJ: Princeton University Press, 1991)
r.p.a.	random phase approximation
STR	Special theory of relativity

1

Criticism, Logical Empiricism, and the General Theory of Relativity

For the human being in his surrounding world there are many kinds
of praxis, and among them is this peculiar and historically late one,
theoretical praxis.

Edmund Husserl[1]

1.1. TWO ACCOUNTS OF PHYSICAL THEORIES

Alfred Tarski's *Introduction to Logic and to the Methodology of Deductive
Sciences* opens with this assertion: 'Every scientific theory is a system of
sentences which are accepted as true and which may be called laws or asserted
statements or, for short, simply statements' (1965: 3).

The 'statement view of theories', as we may call it, was shared by the logical
empiricists, and went almost unchallenged in the philosophy of science
during the middle decades of this century.[2] If, as Tarski does in the case of
mathematics, we add the requirement that the statements of a theory are to
be laid out as a formal deductive system, we get the following thesis:

The sciences are properly expounded in formal axiomatized systems. ... The sciences
are to be axiomatized: that is to say, the body of truth that each defines is to
be exhibited as a sequence of theorems inferred from a few basic postulates or
axioms. And the axiomatization is to be formalized: that is to say, its sentences are
to be formulated within a well-defined language, and its arguments are to proceed
according to a precisely and explicitly set of logical rules.   (Barnes 1975: xi)

The statement view has a long and distinguished history, going back more
than two millennia. In fact, the passage just quoted is not a summary of

---

[1] Husserl [1954] 1970: 111.
[2] One of the few dissenting voices was Stephen Toulmin's [1953] 1960. I discuss his views in
Essay 6.

the views of the logical empiricists, but is Jonathan Barnes' rendering of
'the essential thesis of Book A' of Aristotle's *Posterior Analytics*. His choice
of phrase is only slightly mischievous. Other, more recent, anticipations of
the statement view appear from the seventeenth century onwards in the
methodological writings of, for example, René Descartes, Isaac Newton, Jean
Le Rond d'Alembert, Macquorn Rankine, and Pierre Duhem. I will return
to these figures in later essays. All of them believed that a scientific theory
should be set out *more geometrico*, in the axiomatic manner.

The logical empiricists were also heirs to the investigations into formal
logic and the foundations of mathematics pursued by Gottlob Frege, David
Hilbert, and Bertrand Russell. In 1900, David Hilbert listed the axiomatization of physics among the twenty-three problems facing twentieth-century
mathematicians. Thus, in 1924, when Hans Reichenbach provided a logical
reconstruction of Einstein's special and general theories of relativity, it was
hardly surprising that he should set it out axiomatically.[3] The influence of
Frege and Russell is most obvious in the case of Rudolf Carnap, who had
been Frege's student at Jena. Their 'new logic' (Carnap's phrase)[4] is deployed
in his prescription for the way an ideal physical theory should be articulated.
According to Carnap (1939), a proper philosophical understanding of a
physical theory requires that the theory be set out as a sequence of statements
in the formal language of first-order logic.[5] Each statement in the sequence
must be either an axiom of the system, a mathematical theorem, or deducible
from earlier statements in the sequence using the rules of inference sanctioned
by the logic.[6] These syntactically defined rules are to be supplemented by a
system of semantical rules for the interpretation of the terms of the formal
language. While acknowledging that the concepts of physics are of different
'degrees of abstractness', Carnap divides the terms corresponding to them
into just two classes: elementary terms and abstract terms. The function
of some of the semantical rules is to interpret the elementary terms as
denoting observable or measurable properties of things; the function of the

----

[3] Indeed, Reichenbach's introduction [1924] 1969: 3–4 makes explicit reference to Hilbert's
analysis of axiomatization, as it appeared in the latter's *Foundations of Geometry* [1902] 1921.

[4] Rudolf Carnap's essay, 'Die alte und die neue Logik' appeared in the first volume of *Erkenntnis*
(1930–1). For an English translation see Ayer 1959.

[5] A logic is first order if it allows quantification over individuals, but not over properties. A
first-order language that treats 'is defective' as a predicate cannot express the second quantification
in the statement, '*All* these light bulbs are defective in *some* way or other'.

[6] Extra premisses are allowed—indeed, they must be used—when the theory is used to make a
specific prediction. See Carnap 1939: §§ 19 and 23.

others, to define the abstract terms in terms of the elementary terms, or vice versa.[7]

Carnap's account of scientific theories, modified here and there by Ernest Nagel (1961: chap. 5) and Carl Hempel (1966: chap. 6), became widely accepted, so widely, in fact, that it was referred to as 'the standard view' or 'the orthodox view' by those who were sympathetic to it, and as 'the received view' by those who were not.[8]

A different proposal was made by Patrick Suppes in the late 1960s, and taken up in the next decade by Joseph Sneed and Wolfgang Stegmüller. Following Stegmüller (1979), I will call this approach the 'structuralist view of theories'.[9] Like the logical empiricists, Suppes emphasized the axiomatic method, but he differed from them, first, by using this method to define set-theoretic structures and, secondly, by allowing informal set-theoretic methods of proof where the empiricists had insisted on a rigid adherence to first-order logic. Instead of looking to Hilbert's work in mathematics as a model, Suppes and those who followed him looked to the group of French mathematicians known collectively as Bourbaki.

An example will give the flavour of this approach. Sneed (1979), following Suppes (1957: chap. 12.5), presents a set-theoretic characterization of the systems dealt with by classical particle mechanics. Such a system consists of a set (P) of point masses (particles). Each particle has a well-defined mass (m), and is acted on by various forces (f); its position (s) varies

[7] See ibid. §§ 23–5; note that this account appears at the end of his monograph *Foundations of Logic and Mathematics*.

[8] Hempel's paper 'On the "Standard Conception" of Scientific Theories' appeared in 1970, in the same volume as Herbert Feigl's 'The "Orthodox" View of Theories: Remarks in Defense as well as Critique'. The phrase 'the received view' was coined by Hilary Putnam (1962: 260). For a historical account of the development of the view and the criticisms levelled at it, see Frederick Suppe's 'The Search for Philosophical Understanding of Scientific Theories', written as an editorial introduction to Suppe 1977a.

[9] The terms that have been used by philosophers of science including myself to classify approaches to scientific theories have often been misleading. Carnap's view is sometimes called 'the syntactic view', despite the important role he assigns to semantical rules. The Suppes–Sneed–Stegmüller approach is often lumped together with others (due to Bas van Fraassen 1980, Ronald Giere 1984, and Frederick Suppe 1989) as 'the semantic view', despite the fact that almost the only things all these authors have in common are (a) a frequent use of the term 'model' (see n. 12 below), and (b) a desire to steer the philosophy of physics away from a preoccupation with language—a desire which makes the term 'the semantic view' a very odd choice indeed. For a historical account of the emergence of these various 'semantic views', see the first chapter of Suppe 1989; for a brief analysis of the differences between them, see Hughes 1996 or Morrison and Morgan 1999: chap. 1. In this book I forswear the use of the terms 'syntactic view' and 'semantic view'. Even so, I do not escape controversy. Moulines declares, with emphasis (1998: 81): 'Patrick Suppes ... is *not* a structuralist'.

with time (T). (More formally: Sneed defines *is-a-particle-mechanics* as a set-theoretic predicate which holds of a certain kind of set, an ordered quintuple $<P, T, \mathbf{s}, m, \mathbf{f}>$, whose five members are also particular kinds of sets.) Two clauses of his definition specify that P is a non-empty finite set, and that m is a function from P into the real numbers, such that, for each member p of P, $m(p) > 0$. For Sneed, a *realization* of this structure is a set (call it 'P^*') of particles, such that the mass m^* of each particle is positive and never varies. The clauses that define T, $\mathbf{s}$, and $\mathbf{f}$ ensure that, in any realization, at all times during the time interval T^* both the position $\mathbf{s}^*$ and the acceleration of each particle are well defined, as is the vector sum of the forces $\mathbf{f}^*$ acting on it. By adding to his definition clauses expressing Newton's second and third laws, Sneed places further constraints on the structure, and in so doing defines a *Newtonian classical particle dynamics*.

Along with Wolfgang Balzer and Ulises Moulines, Sneed later showed the versatility of the structuralist approach by applying it, not only to classical particle dynamics, classical collision dynamics, and relativistic collision dynamics, but also to decision theory, Daltonian stoichemistry, simple equilibrium thermodynamics, Lagrangian mechanics, and pure exchange economics (Balzer et al. 1987). The same authors also apply the set-theoretic approach to a general theory of theories, as does Stegmüller (1976: 1979).

In retrospect, we can see that, different though they were, the 'standard conception of theories' and the structuralist account were both guided by the same methodological assumptions. Both parties took the proper objects of philosophical enquiry to be *scientific theories*. Both assumed that the proper way to conduct this enquiry was to provide reconstructions of scientific theories in canonical form. Both agreed with Suppes (1967: 67) that 'the methods and concepts of modern logic provide a satisfactory and powerful set of tools for analysing the detailed structure of scientific theories'. Underlying all of these was the further assumption implicit in the phrase 'scientific theory', the assumption that a common analysis could be applied to all the empirical sciences, from physics to psychology.

These are not the assumptions that inform the essays in this book. There may be useful things to be said about scientific theories in general, but I do not attempt to say them. Such theories as I discuss are theories of physics. In these discussions, however, a theory is not hypostasized as an abstract object, set-theoretic or otherwise, but described in terms of its role in theoretical practice; that is, as one element in a cluster of interlocking theoretical techniques. My aim is not to provide a reconstruction of this practice, but rather to show how it is exhibited in specific examples from the recent and not-so-recent history of physics, and to draw philosophical lessons from doing so. To this end I deploy the resources of modern logic only when it seems useful to do so, which is to say, very seldom.[10]

1.2. PHILOSOPHY OF PHYSICS AS CRITICISM

I shall examine the theoretical practices of physics as they appear in physics journals, treatises, monographs, and the like. I treat these publications as texts; and thereby cast the philosopher of science in the role of critic. Criticism may take a variety of forms. In some hands: 'Criticism is a reformative apparatus, scourging deviation and repressing the transgressive, yet this juridical technology is deployed in the name of a certain historical emancipation' (Eagleton 1984: 12). Terry Eagleton is here describing eighteenth-century French criticism, as practised by Voltaire and Rousseau, but a similar description could be applied to the twentieth-century practices of the logical empiricists. For examples of the scourging of deviance we may look to Reichenbach's dismissal of Hegel in chapter 1 of *The Rise of Scientific Philosophy* ([1951] 1964: 3–4) and Hempel's castigation of the seventeenth-century astronomer Francesco Sizi in his *Philosophy of Natural Science* (1966: 48);[11] for a statement of emancipatory intent, to Herbert Feigl's response to critics of the empiricists' programme: 'There is nothing dogmatic or ritualistic in our movement. It is *not* a religion. Quite to the contrary, it is a reaction against and emancipation from the bondage of metaphysical dogma and speculation' (Feigl 1956: 4).

The logical empiricists recognized just two ways of examining scientific theorizing, both analytic in mode. Famously, Reichenbach distinguished between 'analyses in the context of discovery' and 'analyses in the context of justification'. Feigl draws the distinction thus:

It is one thing to retrace the historical origins, the psychological genesis and development, the social-political-economic conditions for the acceptance or rejection

---

[10] There is nothing original in my renunciation of these assumptions. Two important examples of works in which they play no part—one by a physicist, the other by a philosopher—are John Ziman's *Reliable Knowledge* (1978) and Nancy Cartwright's *How the Laws of Physics Lie* (1983). And, as long ago as 1974, Hilary Putnam emphasized 'the primacy of practice' ([1974] 1981: 78). After three decades, however, Ziman's book remains one of the few available that offers an extended analysis of theoretical practice.

[11] Sizi's offence, committed in *Dianoia Astronomica* (1610), was to employ the Renaissance doctrine of analogy in arguing against Galileo.

of scientific theories; and it is quite another to provide a logical reconstruction of the conceptual structures and of the testing of scientific theories.   (1970: 4)

Neither of these descriptions fits the project undertaken in this book. In examining specific examples of theoretical practice I will pay due attention to their historical context, but my aims are not those of a historian, a psychologist, or a sociologist. And, as will appear later, the assumptions present in Feigl's account of analyses of the second type are precisely those I identified and rejected in the first section of this essay.

Instead, the kind of philosophical assessment I will present resembles a currently unfashionable—even 'pre-structuralist'—type of literary criticism. Q. D. Leavis once wrote, apropos of *Wuthering Heights*: 'The difficulty of knowing that something is a classic is nothing compared to the difficulty of establishing *what kind* of a classic it is—what is in fact the nature of its success, what kind of creation it represents' (1969: 85; emphasis in the original). Analogously, my examination of the theoretical practices of physics is guided throughout by the question: Given that these practices are successful, what is the nature of their success?

In the final Query of the *Opticks* we find Isaac Newton responding to just this question, as it applied to his own theory of universal gravitation. His answer is well known: '[T]o derive two or three general Principles of Motion from Phaenomena, and afterwards to tell us how the Properties and Actions of all corporeal Things follow from those manifest Principles, would be a very great step in Philosophy, though the Causes of those Principles were not yet discover'd' ([1730] 1952: 401–2). This is a descriptive answer, at two levels. One is the level of physical theory. Newton suggests that a mathematical description of the phenomena studied, at the same time general and economical, constitutes an adequate theoretical treatment of them. The other is the level of meta-theory. Newton starts from a description of what his theoretical practice achieves, not what it ought to achieve. 'Descriptive' is to be contrasted, in the first case with 'causal', in the second with 'prescriptive'. Descriptive and prescriptive modes coexist in many philosophical writings about science. Indeed, two pages after the passage I have just quoted, Newton asserts that, 'As in Mathematicks, so in Natural Philosophy, the Investigation of difficult Things by the Method of Analysis, *ought ever* to precede the method of Composition' ([1730] 1952: 404; my emphasis).

Similarly, in Pierre Duhem's *The Aim and Structure of Physical Theory* the prescriptive mode of Part I ('The Aim of Physical Theory') is in marked contrast to the descriptive mode of Part II ('The Structure of

Physical Theory'). A twentieth-century work in which the prescriptive mode predominates is P. W. Bridgman's *The Logic of Modern Physics*, in its demand that each physical magnitude appearing in a theory be given an operational definition. Examples of authors in whose writings the descriptive mode predominates, and with whom my sympathies lie, are contemporary philosophers of physics like Nancy Cartwright, Ronald Giere, and Margaret Morrison.

No work of criticism stands outside time. Each is written in a context that includes the recent history of the field under discussion, be it literature, science, or criticism itself. Newton added his last Questions to the *Opticks* in 1717, in a context supplied by his own achievements and by the criticisms made by his Cartesian detractors. The first edition of Duhem's *The Aim and Structure of Physical Theory* was published in 1906, and the book can be read as a generalization of lessons learned from forty years of disputes over the proper methodology to adopt in theorizing about the electromagnetic field. A decade and a half later, Moritz Schlick, Carnap, and Reichenbach, the founders of logical empiricism, were deeply influenced by Einstein's general theory of relativity. And, more recently, the emphasis placed on models in the last two decades by Cartwright, Giere, and Morrison reflects their widespread use in contemporary physics.[12]

Sometimes this can be a limitation. Bridgman's operationalism, drawing its lessons from the special theory of relativity, prescribed the way to win the battles of a war long over. That said, philosophy of science, like good literary criticism, need not be confined to the work of the immediate past. Q. D. Leavis wrote about *Wuthering Heights* in 1969, and so shared neither the assumptions nor the sensibility of those who read it in 1847 when it first

---

[12] Here Cartwright, Giere, and Morrison differ from the structuralists (see n. 9 above). The set-theoretical entities used by the structuralists are also referred to as 'models', and Patrick Suppes (1960: 289) argues that 'the meaning of the concept of model is the same in mathematics as in the empirical sciences', and that 'the differences to be found in these disciplines is to be found in their use of the concept'. That may be. And, indeed, physicists might choose to model the space-time of special relativity theory, for example, as a set-theoretical structure of the kind Suppes favours. But, such cases are rare. While 'the meaning of the concept of model' may be the same for the physicist as for the mathematician, nevertheless, the type of model the structuralist uses is not part of the physicist's stock in trade. Even if it were, that would not affect the point at issue; Cartwright, Giere, and Morrison are interested in models because physicists use them, Suppes, Sneed, and Stegmüller because they offer a way to reconstruct physical theories. Yet, another approach to theories in terms of models is the *state–space* approach used by, for example, Beltrametti and Cassinelli (1981), Hughes (1989), and van Fraassen (1991), in discussing quantum mechanics. Again, although this approach is quite close to that employed in theoretical practice, that is not the reason why these authors use it. Rather, they use it as a way to understand the internal economy of the theory.

appeared (for one of whom it was 'too odiously and abominably pagan to be palatable to the most vitiated class of English readers'[13]). Yet, her position as a critic is not thereby undermined. The historical and cultural distance that separates her from those readers does not make her insights into the novel any less penetrating. Indeed, one mark of a classic is its capacity to offer a plurality of readings to differently situated readers.[14] The Bard of Avon can become, in Jan Kott's phrase, 'Shakespeare Our Contemporary', and his *King Lear* can be read as Beckett's *Endgame*. Likewise, the insights of contemporary philosophers of science, though achieved by an examination of twentieth-century physics, can nevertheless be applied to the physics of a remoter age. In Essay 3, for example, I shall argue that Ronald Giere's way of characterizing physical theories can be used to provide a reading of Newton's *Principia* that is in many ways more illuminating than those offered a century ago by Ernst Mach and Henri Poincaré.

1.3. LOGICAL EMPIRICISM AND THE GENERAL THEORY OF RELATIVITY

I noted earlier the profound influence of Einstein's general theory of relativity (GTR) on the logical empiricists. Given the centrality of logical empiricism within twentieth-century philosophy of science, this influence is worth tracing in some detail. Its nature and extent are only now being recognized.[15] Einstein published his general theory of relativity in 1915, and during the next ten years Schlick, Reichenbach, and Carnap all provided philosophical analyses of Einstein's theory,[16] and all were agreed that 'Einstein's work, proceeding

---

[13] This assessment, from the *Quarterly Review*, is quoted by Leavis (1969: 65).

[14] See, for example, Kermode 1983: chap. 4.

[15] With the 1991 centennial of the birth of Carnap and of Reichenbach, this omission has been remedied. See, in particular, the papers by Michael Friedman and Don Howard in Salmon and Wolters 1994, together with the commentaries on them by Joëlle Proust and Thomas Oberdan; also the papers by Howard and T. J. Ryckman in Giere and Richardson 1996. Don Howard writes (1994: 46): '[The general theory of relativity] was more than just an exemplar of good science, to which a new philosophy of scientific method would have to be adequated. ... [T]his early engagement with relativity also determined in crucial ways what might be termed the fine structure of emergent logical empiricism'. As Giere has pointed out (1996: 337), one reason why this influence was not recognized earlier was that not until logical empiricism had already become the dominant position in American philosophy of science were any English translations available of Carnap's and Reichenbach's work in the 1920s. None of Reichenbach's three major works in relativity theory appeared in English during his lifetime. Originally published in 1920, 1924, and 1928, their translations appeared in 1965, 1969, and 1957, respectively.

[16] Note in particular Schlick 1915 and 1917; Reichenbach 1920 and 1924; and Carnap 1922.

from an analysis of the meaning of statements about time and space, was actually a philosophical achievement' (Schlick [1930–1] 1959: 58). The question was: In what sense is GTR an empirical theory? (compare: What is the nature of its success?). Schlick, Reichenbach, and Carnap had to mount a philosophical defence of the theory on two fronts. For Machian positivists and strict neo-Kantians alike, the theory was deeply suspect. For the Machian positivist, its fundamental concepts were too remote from sensory experience for it to be a genuinely empirical theory.[17] For the strict neo-Kantian, on the other hand, space and time were, by their very nature, outside the ambit of empirical science. Judgements concerning them were all synthetic a priori, non-analytic yet apodeictically true.

I emphasize the phrase 'strict neo-Kantian'. Some neo-Kantians, notably Ernst Cassirer in Marburg, recognized the achievement represented by Einstein's general theory, and sought to accommodate it within a recognizably Kantian framework. Cassirer suggested ([1921] 1953) that while certain features of space and time (the metric, for example) might be a matter for empirical investigation, there would nevertheless remain, underlying the procedures and interpretation of these investigations, a cluster of tenets about its topology that would have to be acknowledged as synthetic a priori principles.

Schlick, Reichenbach, and Carnap all took a further step away from Kant's original doctrine by making the a priori a matter of convention. Although their views differed in significant ways,[18] common to all three philosophers was the view that the principles of a physical theory can be divided into two groups: foundational principles (as we may call them) and empirical laws. Thus far they followed Kant. They differed from Kant, however, in two respects. In the first place, while they regarded the foundational principles as a priori, they did not view them as apodeictically true; rather, they were chosen by the theorist in order to meet certain meta-theoretic desiderata.[19] In the second place—at least in Reichenbach ([1920] 1965)—only when these foundational principles were laid down did the empirical principles acquire empirical content.

---

[17] Mach himself died in 1916, but his comments on the earlier special theory had been dismissive. For an empiriocritical assessment influenced as much by Avenarius as by Mach, see Joseph Petzoldt 1921.

[18] For details, see Friedman 1994 and Howard 1994, and the accompanying comments by Proust and Oberdan.

[19] Conventionalism has its roots in Heinrich Hertz's Introduction to his *Principles of Mechanics*, which first appeared in 1894, and in the writings of Hermann von Helmholtz and Henri Poincaré. The *locus classicus* is Poincaré [1902] 1952: chaps 4 and 5.

Reichenbach did more than give a programmatic analysis of Einstein's theories. In 1924, he spelled out how the analysis could be synthesized by publishing his axiomatization of the special and general theories of relativity ([1924] 1969). In the Preface he tells us: 'Its goal is the construction of the relativistic theory of space and time with epistemological rigor. The *axiomatic method* is the only method that will reveal the logical structure of the theory with perfect clarity' (ibid. xii–xiii; emphasis in the original).

Within the axiomatization, Reichenbach distinguishes, as is usual, between axioms and definitions. But, his axioms deal, not with mathematical objects like points and lines, but with (idealized) processes and objects like signals and rigid rods. These axioms are empirical propositions, expressing what Reichenbach calls *elementary facts*, 'facts whose interpretation can be derived from certain experiments by means of simple theoretical considerations'. In particular, '[A]ll are facts that can be tested by means of pre-relativistic optics and mechanics' (ibid. 7).[20] Similarly:

[D]efinitions in physics are different from definitions in mathematics. The mathematical definition is a *conceptual definition*, that is, it clarifies the meaning of a concept by means of other concepts. The physical definition takes the meaning of the concept for granted and coordinates it to a physical thing; it is a *coordinative definition*.   (ibid. 8)

Largely because Reichenbach himself provided extended commentaries on it,[21] the best known of these coordinative definitions is the definition of 'Einstein simultaneity' (ibid. 44):

Definition 8. Let a first signal [i.e. a light ray] be sent from A at the time t_1; reflected at B, let it return to A at the time t_3. Then its time of arrival at B will have the following value:

$$t_2 = \tfrac{1}{2}(t_1 + t_3).$$

The purpose of the definition is to provide for the synchronization of spatially separated clocks. The factor '$\tfrac{1}{2}$' appearing in it represents a decision to set the speed of light from A to B equal to its speed from B to A. A *decision*, because

---

[20] There are a few exceptions, he tells us, which 'have been conceived as "thought experiments"' (Reichenbach [1924] 1969: 7).

[21] Reichenbach chooses this example when he describes the nature of coordinative definitions in his Introduction to the *Axiomatization* [1924] 1969: 9–11; he gives a fuller account in Reichenbach [1928] 1957: 123–47.

without the assumption that distant clocks can be synchronized we cannot measure the one-way speed of light, and so the only available measurements of its speed are round trip measurements. Reichenbach argues that, prior to making that decision one can only define what he calls 'simultaneity within the interval' (ibid. 35):

Definition 2. Let a first signal sent from O at time t_1 and reflected at P return to O at time t_2. Then the instant of its arrival at P is to receive the time value

$$t_2 = t_1 + \varepsilon \, (t_3 - t_1 0) < \varepsilon < 1$$

where ε is an arbitrary factor that must have the same value for all points P.

He calls this arbitrariness 'the epistemological relativity of simultaneity' (ibid. 11). Only by *fiat*, by selecting a value for ε, can we make the coordination of the two timescales unique.

Reichenbach generalizes the lesson to be learned from this pair of definitions. In his Introduction to the *Axiomatization* he writes:

This example serves well to point out the arbitrariness of the coordinative definition. ... Just like conceptual definitions, coordinative definitions are arbitrary, and 'truth' and 'falsehood' are not applicable to them. Certain requirements may be imposed on the definition of simultaneity, for instance, that the time values on the clocks should satisfy certain requirements of causality ...; the selection of the coordinative definition will then be restricted. Which requirements are necessary in order to make the definition *unique* cannot be established a priori; these requirements can only be derived from the facts laid down in the axioms. ... Furthermore, it cannot be known a priori whether the definitions are consistent; this result ... follows only from the axioms. ... It is the task of the theoretician to choose his definitions in such a way that, given the axioms, they lead to simple logical relations.   (ibid. 9)

Thus, the final desideratum that leads Reichenbach to adopt Definition 8 is the *descriptive simplicity* of the resulting theory.[22]

Reichenbach's axiomatization was a remarkable achievement. It established a pattern for philosophical analyses of science for the next half-century. It is essentially the paradigm that Feigl looks back to in 1970 when he declares (1970: 4) that the aim of philosophy of science is to provide 'a logical reconstruction of the conceptual structures and testing of scientific

---

[22] For a recent discussion of the conventionality of the simultaneity relations in STR and GTR, see Friedman 1983: iv. 7 and vii. 4.

theories'.[23] Many elements of this paradigm are familiar. The object of Reichenbach's investigation is a theory of physics, and the aim of his analysis is to deduce (in the Kantian sense) the theory's claims to empirical legitimacy. To do so, Reichenbach provides, in his own words ([1924] 1969: xii), 'a construction of the relativistic theory of space-time with epistemological rigor'. Note, however, that his (re)construction is not 'logical' in the narrow sense; nowhere is the apparatus of modern symbolic logic brought into play.

Given the nature and history of the special and general theories of relativity, their use as exemplary cases of scientific theorizing was both inevitable and unfortunate. Inevitable, because they were revolutionary theories of the structure of space and time, and as such were of immediate interest to philosophers. Unfortunate, for much the same reasons. Consider the theories themselves, and their history. First, by definition, revolutionary science is not normal science, and most science—including most theoretical science—is normal science. Secondly, Einstein's theories turned out to be axiomatizable, and—though Hilbert would have wished it otherwise—in twentieth-century physics axiomatizability is the exception rather than the rule.[24] It is true that textbooks of quantum mechanics, for example, often isolate certain propositions and dub them 'axioms of quantum theory'. But, as Bas van Fraassen has pointed out (1980: 65), these propositions typically form 'a fairly straightforward description of a family of models, plus an indication of what are to be taken as empirical substructures of these models'. He continues, 'To think that this theory is here presented axiomatically in the sense that Hilbert presented Euclidean geometry, or Peano arithmetic, in axiomatic form, seems to me simply a mistake'. I agree. And, in other flourishing branches of physics, condensed matter physics for one, axiomatizability is not to be hoped for. Instead, physicists turn to simple models of the systems and phenomena that they study. (One such model, due to Ernst Ising, is the topic of Essay 6.) Thirdly, half a century after its publication, Einstein's theory still

[23] Two qualifications are in order. (a) Then and now criticisms have been levelled both at specific failings in Reichenbach's *Axiomatization* and at his overall project. See, for example, Ryckman 1996. (b) In his treatment of general relativity Reichenbach uses some moderately advanced mathematics, namely the tensor calculus. Consequently, comparatively few professional philosophers have worked their way through it. Neither of these facts, however, contradicts my assertion that the *Axiomatization* was held up as a paradigmatic achievement by the logical empiricists and their analytic successors. Note also that the original logical empiricists were well equipped to work their way through it, even if some of their analytic successors were not.   [24] See Essay 3.

had not become the kernel of a set of theoretical practices—nor of a set of experimental practices, for that matter.[25] Again, as will appear in Essay 2, the contrast with quantum mechanics is striking.

These three considerations are variations on a single theme: that Einstein's theories are too atypical to be used as exemplars for a comprehensive philosophy of physics. Only the second of them, however, the fact that they are axiomatizable, played a significant part in shaping the methodology adopted by the logical empiricists. The major reason why, for good or ill, their methodology took the form it did lies elsewhere. It derives from the fact that Einstein's theories deal with the structure of space and time, topics that for over two millennia have been the subject of philosophical speculation. (Recall Schlick's assertion that 'Einstein's work, proceeding from an analysis of the meaning of statements about time and space, was actually a philosophical achievement'.) In particular, in early twentieth-century Germany and Austria there already existed a philosophical problematic inherited from Immanuel Kant, within which these topics were discussed, and to which Schlick, Reichenbach, and Carnap all responded.[26] Within this problematic, the object of study (for Kant, our knowledge of space and time; for Reichenbach, Einstein's theories) is to be atomized into individual items (for Kant, judgements; for Reichenbach, statements), which are classifiable according to their epistemic warrant (for Kant, as analytic, synthetic a priori, or synthetic a posteriori; for Reichenbach, as empirical laws or coordinative definitions). Sometimes the response to Kant is explicit, as when Reichenbach writes ([1920] 1965: 47): '[W]e must characterize the epistemological position of the principles of coordination. They are equivalent to Kant's synthetic a priori judgements.'[27] Elsewhere, particularly in Schlick's writings of the same period, the debt to Kant is implicit in the key role played by judgements (*Urteile*).

---

[25] See Will 1989: 7–8. The celebrated confirmation in 1919 of Einstein's predictions concerning the deflection of light by the sun was not followed up. Concerning the 1919 observations, Clifford Will comments: 'The measurements [of the deflection] were not all that accurate and subsequent attempts to measure the deflection of light were not much better.'

[26] Carnap's dissertation *Der Raum*, for example, was published in a supplementary volume of *Kant-Studien*.

[27] This is not an isolated instance. The quotation comes from Reichenbach's *Relativitätstheorie und Erkenntnis Apriori* ([1920] 1965), whose title itself is telling. Within this work Reichenbach frequently compares his programme to Kant's; he suggests, for example (ibid. 46), that the question of how the principles of coordination can be both unique and consistent 'is equivalent to Kant's question, "How is natural science possible?" '.

By the early 1930s, the 'linguistic turn' has occurred,[28] and judgements have been supplanted by statements. None the less, elements of the neo-Kantian problematic linger on. Epistemic warrants now function as criteria of meaning. More fundamentally, the logical empiricists' programme remains resolutely atomistic; in the next decades individual statements become the building blocks of logical reconstructions of scientific theories, scientific predictions, scientific confirmations, and scientific explanations.[29] And, no sooner is philosophy of science freed from a reliance on dubious mental entities, than it is laced into the emancipatory straitjacket of first-order logic. We have arrived at the 'standard view' of scientific theories with which this essay began.

1.4. CARNAP'S THESIS AND THE PERIHELION OF MERCURY

In 1939, with first-order logic in mind, Carnap wrote (1939: 60): 'Any physical theory, and likewise the whole of physics, can in this way be presented in the form of an interpreted system, consisting of a specific calculus (axiom system) and a system of semantic rules for its interpretation.' I will call this assertion 'Carnap's thesis'. Note in particular the phrase, '*and likewise the whole of physics*'. The explicit claim of the thesis concerns the adequacy of a specific formal language, but the presupposition it relies on is the one we have just encountered, that physics in its entirety can be adequately represented as a set of statements.[30] It fell out, however, that the reconstruction of all things scientific in terms of statements gave rise to problems that stubbornly resisted resolution.[31] For Bas van Fraassen (1980: 56), these problems were 'purely self-generated ... and philosophically irrelevant', the result of a decision 'to think of scientific theories in a language-oriented way' (ibid. 64). 'The main lesson of twentieth-century philosophy of science', he suggests (ibid. 56), 'may be this: no concept which is essentially language-dependent has any

[28] See Schlick's 'The Turning Point in Philosophy' ([1930–1] 1959), which first appeared in the first volume of *Erkenntnis* (1930–1).

[29] Dudley Shapere 1966: 41–4 provides a perceptive analysis of the senses in which, for the logical empiricists, philosophy of science was equated with 'the logic of science'.

[30] Elsewhere Carnap reiterated and even generalized this claim (1937: 151): 'According to the thesis of *Physicalism* ... all terms of science, including psychology and the social sciences, can be reduced to terms of the physical language. ... For anyone who takes the point of view of Physicalism, it follows that our Language II forms a complete syntactic framework for science.'

[31] See Frederick Suppe's introductory essay to Suppe 1989.

philosophical importance at all'. This last assertion is open to challenge. For example, I take the concepts of a *pidgin* and a *creole* to be 'essentially language-dependent', but Peter Galison (1997) has shown how valuable they are in analysing how theorists and experimenters communicate with each other, a topic that is surely of philosophical interest. I note also that the examples by which van Fraassen buttresses his (very general) claim are all couched in the kind of restricted language that Carnap advocates—that is, a language translatable without remainder into the language of first-order logic. One may agree with van Fraassen that the 'statement view' gives an inadequate account of scientific theorizing, while insisting that philosophers of science should have paid more attention, rather than less, to the actual language (or languages) of science. While the major figures in the logical empiricist movement (Reichenbach, Carnap, Hempel, Nagel) knew a great deal about physics, they were highly selective in what they attended to. They investigated certain aspects of the science very carefully; others they simply ignored. Absorbed in the task of providing logical reconstructions of scientific discourse, they neglected to study the speech genres that scientists actually employ. In brief, in their enthusiasm for 'the new logic', they forgot to be empiricists.[32]

In the last thirty years, philosophers have shown that the theoretical practices of physics are much more intricate—and much more interesting—than Carnap's thesis allows; furthermore, that more perspicuous analyses of these practices are available. For illustration, where better to look than the general theory of relativity—this time examining it, not as it might be reconstructed, but as it was employed on the first occasion of its use?

Einstein laid out the mathematical foundations of his general theory of relativity in three papers, all of them published in the *Proceedings of the Royal Prussian Academy of the Sciences* during November 1915.[33] But, between the second and third of these papers he published another (also in the *Proceedings*)

---

[32] This criticism, of course, is not new. As early as 1966, logical empiricism had become the *ancien régime* of philosophy of science. Dudley Shapere writes (1966: 41): 'In the past decade, a revolution—or at least a rebellion—has occurred in the philosophy of science.' He goes on to provide a succinct analysis and assessment of the logical empiricists' programme and its implementation, and to pronounce an unflattering verdict: 'In their involvement with logical details often without more than cursory discussion of any application to science at all, in their claim to be talking only about thoroughly scientific theories if there are any such, and in their failure or refusal to attend to questions about the historical development of actual science, logical empiricists have certainly laid themselves open to the criticism of being, despite their professed empiricism, too rationalistic in failing to keep an attentive eye on the facts which constitute the subject matter of the philosophy of science' (ibid. 42).

[33] The papers appear as Documents 21, 22, and 25 in *The Collected Papers of Albert Einstein*, vi. (1996). Einstein's paper on the perihelion of Mercury is Document 24 in the same volume.

in which he used the theory to account for the anomalous advance of the perihelion of Mercury. The work of Le Verrier in the nineteenth century had established that the major axis of the planet's orbit of Mercury is not fixed in space, but moves forward at a rate of about 10 minutes of arc per century. The presently accepted rate is about $574''$ (seconds of arc) per century, of which $531''$ can be accounted for by Newtonian theory as the effect of gravitational forces exerted by other planets. The 'anomalous' advance is the $43''$ left unaccounted for.[34] Einstein's analysis of the problem ignores the contributions to the advance of the perihelion that can be traced to other planets. Instead, he shows that GTR can explain the anomalous part of the advance as a gravitational effect due to the Sun alone.

The title of Einstein's paper is 'Explanation of the Perihelion Motion of Mercury by the General Theory of Relativity'. For the logical empiricists, the canonical form of a scientific explanation was the Deductive-Nomological model proposed by Carl Hempel and Paul Oppenheim (1948). On the D-N account, a scientific explanation is a set of statements (of course), set out as an argument in which the premises constitute the *explanans* and the conclusion is the *explanandum*. The premises are of two kinds: some of them are 'general laws' and the others are statements of 'antecedent conditions'; the conclusion is to be derived from the premises by 'logical deduction'. In a further analysis of these components, Hempel and Oppenheim look to 'a formalized model language', the language of first-order logic. As this description suggests, Hempel and Oppenheim's concerns are predominantly linguistic. 'Our problem of analyzing the notion of law', they write ([1948] 1965: 265), 'thus reduces to that of explicating the concept of lawlike sentence'. Setting these internal issues aside, however, we may ask the question: Does the D-N account adequately represent the explanation that Einstein provides?[35]

I will present Einstein's argument in some detail. His paper is divided into two sections: §1 is devoted to the gravitational field produced by a point mass, §2 to planetary motions. Each section begins with a set of equations, referred to as the 'field equations' and the 'equations of motion', respectively. The field equations stipulate relations that hold among the components $\Gamma^{\sigma}{}_{\mu\nu}$ of

---

Where I quote that paper my citation will take the form '831/234', where '831' refers to the page in the original *Proceedings* and '234' the page number in *Collected Papers*, vi.

[34] I am using the figures provided by Will 1989: 15. Einstein quoted the accepted value of the anomaly as $45'' \pm 5''$.

[35] That Einstein's paper does indeed constitute an explanation would be hard to deny.

a gravitational field in a vacuum.[36] (The indices σ, μ, ν run from *1* to *4*. *1*, *2*, and *3* refer to the three dimensions of space, *4* to the dimension of time.) The equations of motion (sometimes called the 'geodesic equations') relate the trajectory of a material point within such a field to the field components $\Gamma^\sigma{}_{\mu\nu}$. Both sets of equations appear in the trio of papers that presented the mathematical foundations of GTR.

So far, so good. We may assume that the field equations and equations of motion of GTR fall under 'the concept of lawlike sentence', as the D-N account of explanation requires. But, before the 'antecedent conditions' can be introduced, considerable work needs to be done.

The field equations of §1 are supplemented by equations that define the field components $\Gamma^\sigma{}_{\mu\nu}$ in terms of the metric tensor $g_{\mu\nu}$. In a flat four-dimensional space-time (i.e. an environment in which there are no masses to warp its structure), the values of $g_{\mu\nu}$ are given by a 4×4 matrix familiar to students of the special theory of relativity:

μ	$=$	ν $=$	*1*	*2*	*3*	*4*
		1	-1	0	0	0
		2	0	-1	0	0
		3	0	0	-1	0
		4	0	0	0	$+1$

Einstein modifies these values of $g_{\mu\nu}$ to allow for the gravitational field produced by a point mass at the origin of his coordinate system.[37] Since the components $\Gamma^\sigma{}_{\mu\nu}$ of the gravitational field are defined in terms of the $g_{\mu\nu}$, the modifications made to the latter result in changes in the former. Einstein calculates these changes, using a procedure he describes as 'successive approximation'. His starting point, the 'null approximation', is provided by the flat space-time itself, where all the $\Gamma^\sigma{}_{\mu\nu}$ have value zero. His 'first approximation' is obtained by using the modified $g_{\mu\nu}$ to calculate the resulting

---

[36] I do not assume a familiarity with the mathematical apparatus of GTR. I use the symbols $\Gamma^\sigma{}_{\mu\nu}$ and later, $g_{\mu\nu}$ simply as a convenient shorthand, and follow Einstein in referring to the $\Gamma^\sigma{}_{\mu\nu}$ as 'components of the gravitational field'.

[37] Einstein takes the $g_{\mu\nu}$ for a flat space-time and *a* adds a term $-\alpha x_\mu x_\nu / r^2$ to each $g_{\mu\nu}$ whose μ and ν are each *1*, *2*, or *3*; *b* adds a term $-\alpha/r$ to g_{44}. The remaining matrix elements, i.e., the $g_{\mu4}$ and $g_{4\nu}$, are left unchanged. Here, x_μ and x_ν are spatial coordinates of the point where the $g_{\mu\nu}$ are evaluated, and r is the distance from that point to the origin, i.e. $r = \sqrt{x_1{}^2 + x_2{}^2 + x_3{}^2}$; α is 'a constant determined by the mass of the Sun'. Effectively, α plays the role of the quantity $2GM_S/c^2$ in Newtonian mechanics, where G is the gravitational constant, M_S is the mass of the Sun, and c is the speed of light, which enters the formula to make the units of distance commensurable with those of time.

field components $\Gamma^\sigma{}_{\mu\nu}$. He verifies that they satisfy the field equations—the equations that govern the relations among the $\Gamma^\sigma{}_{\mu\nu}$ to terms of first order. He then makes a small adjustment to this 'first approximation' $\Gamma^\sigma{}_{\mu\nu}$, so that the field equations are satisfied to terms of second order, and so obtains the 'second approximation' $\Gamma^\sigma{}_{\mu\nu}$.[38]

Each of these approximations is brought into play in §2 of the paper. In both cases Einstein uses the GTR equations of motion to determine the trajectory of a material point within the gravitational field produced by a central mass point—that is to say, within the field that the approximations describe. He restricts himself throughout to motions whose velocity is small compared with the speed of light. The analysis falls into two parts. In the first part Einstein shows that, given the restriction, the first approximation of the gravitational field leads immediately to the Newtonian equations of motion for a material body in such a field. Two familiar results are emphasized: The motion of the body conforms to Kepler's first law (the 'equal area' law), and the sum of its kinetic energy and its gravitational potential energy remains constant. In the second (and longer) part, Einstein uses the second approximation of the gravitational field to show where GTR and Newtonian theory diverge. He obtains an expression for the shift ε of the perihelion of a planet during a single orbit in terms of three other orbital elements: the semi-major axis a, the eccentricity e, and the period T. When the values of these elements for Mercury's orbit are inserted, the value obtained for ε corresponds to a rate of advance of the perihelion of $43''$ per century. The anomaly has been explained.

This final move of the argument brings us back to the D-N account of explanation. An abstract of Einstein's paper, written with that account in mind, might run as follows:

The anomalous perihelion shift of Mercury is shown to follow from a set of laws (the field equations and laws of motion of GTR, supplemented by the definitions of the components of the gravitational field in terms of the metric tensor), together

---

[38] A rough and ready example will show the principle underlying the method of successive approximation. Assume that a physical magnitude is theoretically shown to have the value $m1 + c_1a + c_2a^2 + c_3a^3 + \ldots$, where a is small compared with 1. Then, unless the coefficients c_i are large, $c_1a > c_2a^2 > c_3a^3 > \ldots$. To a first approximation the magnitude may be assumed to have the value $m1 + c_1a$, to a second approximation the value $m1 + c_1a + c_2a^2$, and so on. In the present case, in the first approximation, $\Gamma^\sigma{}_{44} = -\alpha x_\sigma/2r^3$, in the second $\Gamma^\sigma{}_{44} = -\alpha x_\sigma/2r^3 + \alpha^2/2r$. These are the only components that play a part in §2 of the paper, and for that reason they are the only 'second approximation' components that Einstein evaluates.

with a set of statements of antecedent conditions (the values of selected elements of Mercury's orbit).

For good measure, one might amplify the abstract by invoking a distinction from Hempel's later writings: 'The field equations, which summarize the relations between the components of the gravitational field, are *internal principles* of GTR; the equations of motion, which relate these theoretical terms to observational terms describing the motions of bodies, are the *bridge principles* of the theory' (1966: 72–5). I will examine each of these proposals in turn.

The abstract could well be called the 'null approximation' of Einstein's argument, true as far as it goes, but remarkable for what it leaves out. The major problem is the squishy phrase 'is shown to follow from'. If the abstract is to conform precisely to the D-N model, the phrase would have to be replaceable by 'is logically deducible from'. But, for two reasons (at least), Einstein's argument cannot be reconstructed in this way. The first is that in §1 Einstein appeals to a key postulate that the abstract ignores; the second, that in §2 he uses the method of 'successive approximation'.

The postulate appears at the very beginning of the argument, where Einstein presents a metric tensor $g_{\mu\nu}$ for the gravitational field produced by a centrally placed point mass. Although he claims (832/235) to derive ('*abzuleiten*') his solution, the claim is not borne out by the procedure he describes. He assumes that the difference between each matrix element $g_{\mu\nu}$ of the required metric tensor and the corresponding element for the flat space-time is small (compared with 1). He then proposes a solution in terms of these differences, and points out (a) that it satisfies certain conditions (for example, that it does not vary with time, that it displays spherical spatial symmetry, and that it becomes a flat space-time at infinity);[39] and (b) that it yields field components $\Gamma^{\sigma}{}_{\mu\nu}$ that satisfy the field equations to magnitudes of first order. No further justification is given.

For the reader working through the paper for the first time, the postulated $g_{\mu\nu}$ seem to have been plucked from the air. But, in the context of the paper as a whole, this postulate is not an independent and unwarranted premiss to be used, along with the field equations and the equations of motion, in the deduction of the *explanandum*. Rather, it receives a retrospective

---

[39] Additional constraints are that $\sqrt{g} = 1$, and that $g_{\mu 4} = 0 = g_{4\mu}$, where μ is the index *1, 2,* or *3*. The full solution was given in n. 4 of his paper. It is easy to see that it conforms to all these conditions provided that α is small.

justification as the argument unfolds. In §1, the postulated $g_{\mu\nu}$ are used to obtain the first approximation $\Gamma^\sigma_{\ \mu\nu}$ for the components of the gravitational field, and these in turn, together with the equations of motion, yield in §2 the Newtonian equations for bodies moving in a centripetal gravitational field. Thus, even before the second approximation for the field is used to predict the 'unexplained residue between observation and Newtonian theory' (839/242), the conjunction of GTR and the postulated metric tensor has satisfied a major *desideratum*: it can recapture the results of Newtonian theory.[40]

I cited the use of the method of 'successive approximation' as the second reason why Einstein's argument cannot be reconstructed as a strictly logical deduction. At every stage, Einstein's argument is built around approximations. Subsections of §1 are headed 'First Approximation' and 'Second Approximation', and the first two pages of §2 contain successively the phrases: 'as a first approximation', 'a first approximation', 'for a first approximation', 'to a more exact order of magnitude', 'to magnitudes of first order exactly', 'to magnitudes of second order exactly', 'to magnitudes of second order exactly' (again), 'to a first approximation', 'exact to magnitudes of second order', 'the exact validity of', and 'exactly, to magnitudes of second order'. Notice that 'exact validity' is rare enough to be remarked on. The logical empiricists, however, say virtually nothing about approximations.[41] The spartan languages of their logical reconstructions do not accommodate sentential operators of the kind I have quoted.[42] Nor can these operators be elevated into the metalanguage, as locutions of the form 'is true to the nth order of approximation', without undermining Carnap's thesis.

[40] It seems likely that Einstein obtained the metric tensor he used by working backwards 'arguing from effects to causes', as Descartes puts it, from the requirement that it should yield Newton's equations at least to first order. But, this would not undermine the pragmatic justification of the postulate. Note that for reasons connected with the covariance of the field equations he explicitly refrains from 'going into the question of whether [the postulated metric tensor] is the only one possible' (832/235). Within two months of the publication of Einstein's paper, an exact solution of the problem was given by Karl Schwarzschild. Because Schwarzschild was serving in the German army on the Russian front, his communication was read to the Royal Prussian Academy by Einstein. Two months later, Schwarzschild died of a disease contracted at the front. See Pais 1982: 255. A version of Schwarzschild's solution is given in Weinberg 1972: 175–80.

[41] The word 'approximation' appears neither in the index of Ernest Nagel's 600-page *The Structure of Science* (1961), nor in the index of Hempel's 500-page *Aspects of Scientific Explanation* (1965). Cf. Cartwright's *How the Laws of Physics Lie* (1983), which discusses approximations on thirty of its 216 pages.

[42] Tellingly, when Carnap (1956: 42) was prepared to tweak the standard view of theories to permit the use of the modal operators $\Box$ *necessarily* ... and $\Diamond$ *possibly* ... , this was done, not because the writings of physicists contained such operators, but because the logical analysis of the concept of a *law* seemed to require it. See Essay 3 for a discussion of the laws of physics.

The second proposal I put forward was that the field equations of §1 and the equations of motion of §2 could be regarded as internal principles and bridge principles of GTR, respectively. Prima facie, there is certainly good evidence for this proposal. All the terms in the field equations are theoretical terms (the field components $\Gamma^{\sigma}{}_{\mu\nu}$ or their first derivatives), whereas the equations of motion include, in addition to the field components, terms like $dx_{\nu}{}^{2}/ds$ and dx_{ν}/ds that are at least 'antecedently understood', in Hempel's phrase, and can be unpacked in terms of the measurable quantities *distance* and *time*. As we have seen, Einstein moves from these equations (and the second approximations of the field components $\Gamma^{\sigma}{}_{\mu\nu}$) to the rate of advance of Mercury's perihelion in two steps. He first obtains an expression for the advance ε of the perihelion in a single orbit (an angle) in terms of orbital elements: the semi-major axis a (a distance), the eccentricity e (a ratio of distances), and the period T (a time). He then plugs in the known values of Mercury's orbital elements into the expression and calculates the rate at which the planet's perihelion advances per century.

There is, however, an equivocation here, a conceptual gap between the formula for the quantity ε as it is obtained in the first step, and the same formula as it is used in the second. The equations of motion, from which the expression for ε is obtained, are introduced as 'the equations of a *material point* in a gravitational field' (35/238; my emphasis). Thus, when the symbols a, e, and T first appear in the expression for ε, they refer to orbital elements of a theoretical idealization, a 'material point'. But, in the next sentence Einstein writes (839/242): 'For the planet Mercury the calculation yields an advance of the perihelion of about $43''$ a century', a calculation in which the same symbols refer to properties of a physical object with a diameter of more than 5,000 kilometres. On their own, the equations of motion do not suffice to bridge the gap between theoretical physics and the physical world.[43]

---

[43] Ernest Nagel (1961: 160) devotes a paragraph to the topic of 'point masses', albeit in a discussion of the laws of classical, rather than relativistic, mechanics. He writes: 'The axioms of motion must be supposed to be formulated for so-called "point masses". ... The application of the axioms of motion to physical bodies, which are clearly not point masses, thus assumes an extension of the fundamental theory to cover the motions of systems of point masses that are subject to more or less rigid material constraints. Such an extension does not require the introduction of any new theoretical ideas.' But, this extension does not do the work of a bridge principle. The axioms still apply to point masses, even if there are many more of them than before and they are now subject to new constraints. Nagel continues: 'The facts just noted make it evident that the axioms of motion are theoretical statements ...; they are not statements about experimentally specified properties' (emphasis in the original).

The obvious—and legitimate—response to this line of thought is to point out that on the scale of the solar system this diameter is very small, less than one ten thousandth of the mean distance of Mercury from the Sun, and so may be assumed negligible. This assumption is implicit in Einstein's argument. In fact, a rereading of his paper discloses a whole cluster of implicit assumptions of this kind. These assumptions emerge very clearly if the argument is presented in an idiom now standard in the philosophy of physics.

Einstein's aim in the paper is to show that the GTR can explain the anomalous precession of the perihelion of Mercury—that is to say, to explain one component of the behaviour of a complex physical system, the solar system. The GTR is applied, however, not to this physical system, but to a theoretical model. The relation of the model to the system is like that of a Hollywood stuntman to a film star: one acts as a stand-in for the other. The model is both simplified and idealized.[44] It is simplified in that only two bodies are represented in it, Mercury and the Sun. The other planets are ignored, even though Newtonian theory tells us that their combined effect is responsible for 90 per cent of the perihelion's advance. The implicit assumption here is that the effects on the perihelion attributable to these planets are causally independent of those attributable to the Sun. The model is idealized in that the Sun is represented by 'a point mass' ('*ein Massenpunkt*') and Mercury by a 'material point' ('*materieller Punkt*').[45] The difference in the descriptions of these idealizations reflects a difference in their roles. The point mass is introduced in §1 as the source of a gravitational field, the material point in §2 as a body whose motion that field determines. There is no reciprocity between them. The mass point is motionless; the material point makes no contribution to the field.[46] We have already noted the assumptions implicit in the representation of Mercury as a material point. In representing

---

[44] I use the terms 'simplification' and 'idealization' where others—for example, Cartwright, Suppe, Morrison, and Martin Jones—prefer to use 'abstraction' and 'idealization'.

[45] In at least two respects, these 'ideal objects' resemble Kantian 'ideals'. They represent the non-empirical limit of an empirical progression; and, if examined closely, they give rise to problems related to infinity. See the 'Antinomies of Pure Reason' in Kant's *Critique of Pure Reason* (1781).

[46] In these respects, Einstein's model is strikingly similar to that used by Newton in the early sections of Book I of the *Principia*, save that Newton does not there attribute the 'centripetal forces' that appear in his theorems to a central mass. That is left to Book III. See, respectively (a) the Scholia to Definition V and to Proposition IV, Theorem IV of Book I; and (b) Proposition V, Theorem V of Book III. In contrast, Einstein indicates from the start that his model will represent the solar system, in §1 by a laconic '*die Sonne*' in parentheses immediately after '*ein Massenpunkt*', and in §2 by its title: '*Die Planetenbewegung*'.

the Sun as a point mass, Einstein assumes that the effect of the Sun's mass on the space-time metric of the region surrounding it would be the same whether that mass was distributed over a volume 1.3 million times the volume of the Earth, or (*per impossibile*) was concentrated at a point.

The lesson to be drawn is this. Einstein faced the problem of applying a foundational theory (GTR) to a physical system (the solar system); to make this problem tractable he needed, not only the equations of the theory, but also a theoretical model of the system. Built into the model was a cluster of implicit assumptions. These, together with the assumption that the theory held for the system in question, added up to what Giere calls the 'theoretical hypothesis', the hypothesis that the model-plus-theory provided an adequate representation of the physical system under scrutiny. 'Adequacy', of course, is a pragmatic virtue, whose nature varies from context to context.[47] The theoretical hypothesis implicit in Einstein's treatment of Mercury's perihelion was that the two-body model he employed, together with the GTR, adequately represented the interaction between the Sun and Mercury mediated by the structure of space-time. In this context, the criterion of adequacy was that the functional dependence of a particular property of the material point in the model (the rate of advance of its perihelion) on other properties (its orbital elements) should accurately reflect a corresponding dependence among the properties of the planet Mercury. Earlier, I suggested that a 'conceptual gap' existed between these two dependencies. The analysis I have given does not close the gap; in fact, it brings it into sharper focus, as an inescapable feature of theoretical practice. In doing so, it explodes the myth that theoretical predictions and explanations can be furnished by strictly logical deductions whose only premises are laws and statements of antecedent conditions, as Carnap's thesis would suggest.

1.5. CODA AND INTRODUCTION

My commentary on Einstein's paper is offered, both for its own sake, and as an illustration of my approach to the philosophy of physics. Throughout this volume I will examine theoretical practice as it appears in individual texts, and attend to theories, not as they may be reconstructed, but

---

[47] Giere cashes out 'adequacy of representation' in terms of 'similarity'. I discuss his choice in Essay 5.

as they are used. Inevitably, any redescription of theoretical practice will be to some degree a reconstruction. For example, I redescribed Einstein's treatment of the anomalous advance of Mercury's perihelion in terms of a simplified and idealized model, although the term 'model' is nowhere to be found in Einstein's paper. None the less, my aim was to make explicit what was implicit in the original text.[48] The problem with a reconstruction is that it is put together with a specific agenda in mind, at the risk of distorting or neglecting much of the original. In his *Axiomatization*, for example, Reichenbach's avowed aim was to reconstruct Einstein's theories with 'epistemic rigour'. But, that begs the question: Was their epistemological status the only aspect of these theories worth investigating?

The phrase 'theoretical practice' might suggest that I undertake 'analyses in the context of discovery'. That is not the case. Contemporary journals of physics do not provide the kind of narrative, full of hopeful beginnings and dead ends, that Johannes Kepler gives us in his *Astronomia Nova*. Tracing the steps that led him to his realization that Mars moves in an elliptical orbit, Kepler wrote: 'Why should I mince my words? The truth of Nature, which I had rejected and chased away, returned by stealth through the backdoor, disguising itself to be accepted. ... Ah, what a foolish bird I have been.'[49]

Such frankness is rarely encountered in the pages of the *Physical Review*.[50] The intellectual narratives that appear in contemporary physics journals are analogous to those nineteenth-century sagas of exploration of the American West in which carefully selected episodes from separate expeditions were stitched together to present a unified 'ideal narrative' to the reader. And, just as a literary or historical scholar may undo the stitching and so display the individual narratives from which the ideal one was constructed, so the philosopher or historian of science can sometimes tease apart the

---

[48] Einstein indicates that his model represents the solar system; see n. 46 above.

[49] Johannes Kepler, *Astronomia Nova*, IV, Cap. 5, 8, quoted and translated by Arthur Koestler (1968: 338). Wilson 1968 suggests that Kepler's account of the path he took to his discoveries was not historically accurate. That does not, however, affect the main point I am making.

[50] Inevitably, Einstein is the exception. On the opening page of the first of his 1915 trio of papers on general relativity he writes (778/215): 'For these reasons I completely lost confidence in the field equations I had set out, and looked for a method of restricting the possibilities in a natural way. Thus I came back again to the demand for the general covariance of the field equations which I had abandoned, but with a heavy heart, three years ago when I worked with my friend Grossmann.'

chronologically separate elements within the composition of a scientific paper.[51] That enterprise, however, is not part of my project.

I give content to the phrase 'theoretical practice' in Essay 2. The essay examines a sequence of four papers by David Bohm and David Pines. Published in the *Physical Review* in the early 1950s, the quartet gives a theoretical account of the behaviour of electrons in metals. Within these papers the authors engage in a practice that includes the use of models, narratives, approximations, and two distinct theoretical manifolds. Each of these elements receives attention in the essay, as does the reception accorded to the quartet. I treat these papers as written utterances that take up and respond to earlier utterances in the genre, and themselves invite responses from their intended readers.

Most of the individual texts I examine are well known, at least to physicists. That is by design. Since my aim in investigating the theoretical practices of physics is to enquire into 'the nature of their success', it behoves me to use as case studies examples which were indeed successful. For Q. D. Leavis, the status of *Wuthering Heights* was not in doubt. It was an established classic. Newton's *Principia*, which I examine in Essay 3, also fits that description, as does Einstein's paper on the Brownian motion, which I examine in Essay 4, but some of my other examples (the Bohm–Pines quartet, for instance) do not. All of them, however, are examples of successful practice in one reasonably clear-cut sense. They were all regarded by the scientific community as notable contributions to the field, and their value was attested, at least in the twentieth-century cases, by the frequency with which they were cited, and by the attention paid to them, first in specialized monographs and then in textbooks.[52] Their success did not mean that they

[51] An example: examination of Niels Bohr's 1913 paper 'On the Constitution of Atoms and Molecules' reveals that his model of the atom was not 'devised in order to explain experimental laws about the line spectra of the various elements', as, for example, Nagel claims (1961: 94). Rather, the discovery that it could do so confirmed Bohr in his belief that his model was an adequate one. See Hughes 1990b: 75–8.

[52] In Essay 2, I document the reception accorded to the quartet of papers I study there. Note that the criteria I have given already eliminate many published contributions to science. A 1973 study showed that 'about one half of all papers that are published in the more than 2100 science journals abstracted in the *Science Citation Index* do not receive a single citation during the year after the paper is published' (Cole and Cole 1973: 228; quoted Ziman 1978: 130 n.). Without knowledge of the typical interval between submission and publication it is not clear just what to make of this statistic. Furthermore, some papers lie dormant, as it were, until rediscovered at a later date. Alfred Wegener's 1912 papers on continental drift, for instance, were ignored until the 1960s. My claim, however, is not that all good papers are successful, but that a criterion, albeit sociological, of success exists.

went unchallenged. In Essay 8, I examine the 1959 paper by Yakir Aharonov and David Bohm in which they predicted the effect that now bears their names. Their proposal met with vigorous opposition from some quarters. *Wuthering Heights*, it may be recalled, was given highly unflattering reviews by Emily Brontë's contemporaries; similarly, while no one described the Aharonov–Bohm proposal as 'odiously and abominably pagan', the A–B effect was dismissed as 'non-existent', and treatments of it as 'vitiated by an erroneous conception of the behaviour of the wave phase', which comes to much the same thing.[53]

I continue the discussion of the A–B effect in Essay 8, by looking at the experiments by Tonomura et al. that—in the eyes of most physicists—conclusively established its existence. That essay and those on either side of it are the only places in the book where experimental practice receives significant attention. This is neither an oversight on my part nor a covert suggestion to the reader that theoretical practice is the only game worth watching. If anything, it is an indication that my philosophical habits owe more to the logical empiricists than I care to acknowledge. Old dogs, old tricks. Be that as it may, I am aware that, in its emphasis on theoretical practice, this collection of essays gives a one-sided picture of physics. Ideally it would be complemented by a volume that concentrated on experimental practice and instrumentation. Luckily, one already exists, written by my colleague Davis Baird.[54]

---

[53] The quotations appear in Bocchieri, Loinger, and Siragusa 1979: 1–2; their paper is entitled 'Non-existence of the Aharonov–Bohm Effect'. Interestingly, S. M. Roy, while using a similar title for another paper, nevertheless refers to the work of Aharonov and Bohm as a 'classic proposal' (Roy 1980: 111).

[54] Davis Baird, *Thing Knowledge* (Berkeley, Calif.: University of California Press, 2004).

2

Theoretical Practice: The Bohm–Pines Quartet

Were a kind of semiotic egalitarianism to direct us to regard as texts the papers that regularly appear in *The Physical Review*, their literary dimension must seem deeply secondary.

Arthur Danto[1]

PREAMBLE

My egalitarian tendencies will be all too evident throughout this essay, dealing as it does with four papers from the *Physical Review*. True, I largely neglect their literary dimension; whatever the attention I pay to the narrative elements in them will betray a sadly undiscriminating taste.

Pace Danto, however plebian they may be, these papers are indisputably texts. They are written utterances that take up and respond to earlier utterances in the genre, and themselves invite responses from their intended readers. This explains why a paper can be too original for its own good, why early papers in chaos theory, for instance, were denied space in physics journals.[2] For, although each paper is individuated by the original contribution it offers, the dialogic relation in which it stands both to its predecessors and its successors requires that all of them be informed by a common set of assumptions. These shared assumptions—some methodological, others theoretical, some explicit, many tacit—provide a normative framework for theoretical discourse. This in turn enables us to speak meaningfully of 'theoretical practice'. The details of the framework may vary from one sub-discipline of physics to another. They will also change with time; under the pressure of theoretical advances some elements of the framework will be

[1] Danto 1986: 136.   [2] See Ruelle 1991.

jettisoned, or fall into disuse, while others become accepted in their place. Thus, one should properly think of physics as involving, not theoretical practice *tout court*, but a set of theoretical practices indexed by sub-discipline and date.

The phrase itself, 'theoretical practice', though not actually oxymoronic, is little used in the philosophy of physics.[3] And, in describing a theoretical advance, physicists will rarely allude to the practices that led to it. For example, in the transcript of one of Richard Feynman's *Lectures on Physics* we read:

It was first observed experimentally in 1936 that electrons with energies of a few hundred to a few thousand electron volts lost energy in jumps when scattered from or going through a thin metal foil. This effect was not understood until 1953 when Bohm and Pines showed that the observations could be explained in terms of the quantum excitations of the plasma oscillations in the metal.   (Feynman, Leighton, and Sands 1963–5: ii. 7.7)

Feynman is here concerned only with the result that Bohm and Pines obtained, not with the strategies they used to obtain it. In his lecture, it appears as one illustration among others of the fact that '[the] natural resonance of a plasma has some interesting effects' (ibid.). But, the work that produced it, the journey rather than the end, offers an illustration of a different kind. It displays an example of the theoretical practices of solid state physics in the mid-twentieth century.

This essay has three parts. In 2.1, I give synopses of the four papers in which Bohm and Pines presented a theoretical account of plasma behaviour; in 2.2, I provide a commentary on these papers; in 2.3, I describe the reception they were accorded; I then offer some remarks concerning theoretical practice and a brief note on the methodologies available to philosophers of science. The commentary in the second (and longest) part contains discussions of the components of theoretical practice exhibited in the quartet. They include models, theoretical manifolds, modes of description, approximations, the relation between theory and experiment, and deduction. Part 2.1 can thus be read as an introduction to the notion of 'theoretical practice' that takes the Bohm–Pines quartet for illustration. Likewise, 2.3 illustrates, amongst other things, how theoretical practice evolves through time.

---

[3] Edmund Husserl's *The Crisis of the European Sciences* is too seldom studied by philosophers of science.

2.1. THE BOHM–PINES QUARTET

2.1.1. Introduction

Between 1950 and 1953, David Bohm and David Pines published a sequence of four papers in the *Physical Review*. They were collectively entitled, 'A Collective Description of Electron Interactions', and individually subtitled, 'I: Magnetic Interactions', 'II: Collective *vs* Individual Particle Aspects of the Interaction', 'III: Coulomb Interactions in a Degenerate Electron Gas', and, 'IV: Electron Interaction in Metals'.[4] In this quartet of papers, Bohm and Pines had two aims: one specific and the other general. The specific aim was to provide a theoretical account of the behaviour of electrons in metals; as their title announces, this account was to be given in terms of a 'collective description of electron interactions' rather than 'the usual individual particle description' (BP I: 625). Their more general aim was to explore, through this analysis, a new approach to many-body problems.

The background to their investigation, roughly sketched, was this. Since the work of Paul Drude at the turn of the twentieth century, the accepted explanation of the high electrical conductivity of metals was that some, at least, of the electrons in the metal were free to move through it. On the account that emerged, the valence electrons in a metal are not attached to specific atoms; instead they form an 'electron gas' within a regular array of positive ions, the 'crystal lattice'. This model was modified by Arnold Sommerfeld (1928). He pointed out that the electrons in a metal must obey the Pauli exclusion principle, according to which no more than one electron in a system can

---

[4] I cite these papers as BP I, BP II, BP III, and P IV. Except for the abstract of BP I, the text on each page is set out in two columns. In my citations, the letters 'a' and 'b' after a page number designate, respectively, the left and right hand columns. The fourth paper was written by Pines alone. When the first paper was published, both authors were at Princeton University, Bohm as a junior faculty member and Pines as a graduate student. Over the period in which they were written Pines moved, first to Pennsylvania State University and then to the University of Illinois. Between the time BP II was received by the *Physical Review* (Sept. 1951) and the time it was published (Jan. 1952), Bohm had moved to the University of São Paolo in Brazil. By refusing to testify about colleagues and associates, Bohm had fallen foul of Joseph McCarthy's Un-American Activities Committee, and was unable to obtain a post in the USA when his contract at Princeton expired. See p. 4 of the introduction to Hiley and Peat 1987. This hiatus in Bohm's life may explain why Pines alone was responsible for the fourth paper. It also partially explains the four-year gap between the first and third. Note, however, that Pines attributes the delay to 'our inability to come up with a simple mathematical proof that [certain] subsidiary conditions could be fulfilled'—conditions which had to be satisfied if their argument was to go through (Pines 1987: 75).

occupy a particular energy level; collectively they need to be treated as a *degenerate electron gas* (hence the title of BP III).[5]

In the 1930s and 40s, a theory of the motion of electrons in a metal was developed using the 'independent electron formulation', otherwise referred to as the 'independent electron model'. On this approach, the electrons are treated individually. To quote John Reitz (1955: 3): '[T]he method may be described by saying that each electron sees, in addition to the potential of the fixed [ionic] charges, only some average potential due to the charge distribution of the other electrons, and moves essentially independently through the system.'

The independent electron theory enjoyed considerable success. The values it predicted for a number of metallic properties (electrical and thermal conductivity among them) agreed well with experiment (Pines 1987: 73). However, it failed badly in one important respect: the predicted cohesive energy of the electrons in the metal was so small that, were it correct, most metallic crystals would disintegrate into their constituent atoms. Furthermore, as Pines puts it (ibid.), '[T]heoretical physicists … could not understand why it worked so well'.[6] The challenge facing Bohm and Pines was to formulate a theory of the behaviour of electrons in a metal that both acknowledged the mutual interactions between electrons and showed why the independent electron model, which ignored them, was so successful.

They looked for guidance to the research on plasmas performed by physicists like Irving Langmuir in the 1920s and 30s.[7] Langmuir used the term 'plasma' to refer to an ionized gas, such as one finds in a fluorescent lamp or in the ionosphere, where electrons are stripped from their parent atoms, in one case by an electrical discharge, in the other by ultraviolet radiation from the Sun. The state of matter that results resembles that in a metal in that a cloud of free electrons surrounds a collection of heavier, positively charged ions.[8] There are differences. The ions in a gas plasma move freely, albeit much more slowly than the electrons, whereas in a metal their motion is confined to thermal vibrations about fixed points in a regular lattice. More importantly,

---

[5] Whereas an ideal gas obeys Maxwell–Boltzmann statistics, as a result of the exclusion principle a degenerate electron gas obeys Fermi–Dirac statistics.

[6] For more details, see Pines 1955: 371–4. An extended discussion of the independent electron theory is given in Reitz 1955.   [7] The paper cited in BP I is Tonks and Langmuir 1929.

[8] Matter also enters the plasma state at the temperatures and pressures associated with thermonuclear reactions. Interest in controlled fusion has prompted much of the research on plasmas since the Bohm–Pines papers were published.

there are about 10^{11} times as many electrons per unit volume in a metal than in a gas plasma. This is the reason why the electron gas in a metal should properly be treated quantum-mechanically, as a degenerate electron gas, whereas the electron gas in a gas plasma may be treated classically (see n. 5).

Despite these differences, Bohm and Pines saw Langmuir's investigations of plasmas as 'offering a clue to a fundamental understanding of the behaviour of electrons in metals' (Pines 1987: 67).[9] They set out to show that the electron gas in a metal would manifest two kinds of behaviour characteristic of gas plasmas: that high frequency 'plasma oscillations' could occur in it, and that the long-range effect of an individual electron's charge would be 'screened out' by the plasma. Paradoxically, both these effects were attributed to the Coulomb forces between pairs of electrons, the electrostatic forces of repulsion that exist between like charges. 'Paradoxically' because, on this account, the plasma screens out precisely those long-range effects of Coulomb forces that are responsible for plasma oscillations. Note, however, that the screening effect, if established, would go some way towards explaining the success of the independent electron theory of metals.

Within the quartet, a mathematical treatment of plasmas is interwoven with an account of its physical significance and a justification of the methods used. Synopses of the four papers are given in the next four sections of this essay. They have been written with two aims in mind: first, to present a general overview of each paper; secondly, to provide more detail about those parts of the quartet that I comment on in Part 2.2 of the essay. The two aims pull in different directions. To alleviate this tension, in each synopsis I distinguish the overview from the more detailed material by enclosing the latter in brackets ('((' and '))'). This allows the reader to skip the bracketed material on a first reading to attend to the overview, and to go back later to browse on the amplified version.

2.1.2. BP I: Magnetic Interactions

The Introduction to BP I is an introduction to the whole quartet. Its first six paragraphs introduce the reader to the independent electron model and its shortcomings, and to the phenomena of screening and plasma oscillations. These phenomena, the authors tell us, occur in an electron

---

[9] A brief summary of previous work along these lines, together with some critical observations on it, is given in BP II: 610b.

gas of high density. In a footnote (BP I: n. 1) they point out that 'The [negatively charged] electron gas must be neutralized by an approximately equal density of positive charge'. In a metal, this charge is carried by the individual ions of the metal. But, 'in practice the positive charge can usually be regarded as immobile relative to the electrons, and for the most applications [*sic*] can also be regarded as smeared out uniformly throughout the system' (ibid.). The presence of this positive charge (whether localized or uniformly distributed) effectively screens out short-range interactions beyond a very small distance. At long range, however, plasma oscillations, a collective phenomenon, can occur. These are longitudinal waves, 'organized oscillations resembling sound waves' (BP I: 625a), in which local motions are parallel to the direction of propagation of the waves. A plasma can also transmit organized transverse oscillations, in which local motions are at right angles to the direction of wave propagation. They may be triggered by an externally applied electromagnetic field, a radio wave passing through the ionosphere, for example. This applied field will produce oscillations of the individual electrons, each of which will in turn give rise to a small periodic disturbance of the field. Only if the cumulative effect of all these small disturbances produces a field in resonance with the original applied field will the oscillations become self-sustaining. As we learn at the end of the Introduction (BP I: 627a), transverse oscillations like these are the main topic of the first paper.

Bohm and Pines treat these oscillations in two ways. In their first treatment they use the techniques of classical physics; in the second those of quantum mechanics. So that the results of the first can be carried over to the second, they use 'Hamiltonian methods'—that is, they describe the system comprising the electrons and the field by its *Hamiltonian*, an expression that specifies its total energy.[10] The authors write (BP I: 626b):

This Hamiltonian may be represented schematically as

$$H_0 = H_{part} + H_{inter} + H_{field}$$

where H_{part} represents the kinetic energy of the electrons, H_{inter} represents the interaction between the electrons and the electromagnetic field, and H_{field} represents the energy contained in the electromagnetic field.

[10] I say more about the Hamiltonian and its significance in sections 2.2.3 and 2.2.4 of this essay.

This Hamiltonian is expressed in terms of the position and momentum coordinates of the individual electrons and field coordinates of the electromagnetic field.[11] Bohm and Pines then introduce a new set of coordinates, the 'collective coordinates',[12] and, as before, a distinction is made between particle coordinates and field coordinates. A mathematical transformation replaces the original Hamilton, H_0, expressed in the old variables, by another, $H_{(1)}$, expressed in the new ones.

((Given various approximations:

[T]he Hamiltonian in the collective description can be represented schematically as

$$H_{(1)} = H^{(1)}{}_{part} + H_{osc} + H_{part\ int}$$

where $H^{(1)}{}_{part}$ corresponds to the kinetic energy in these new coordinates and H_{osc} is a sum of harmonic oscillator terms with frequencies given by the dispersion relation for organized oscillations. $H_{part\ int}$ then corresponds to a screened force between particles, which is large only for distances shorter than the appropriate minimum distance associated with organized oscillations. Thus, we obtain explicitly in Hamiltonian form the effective separation between long-range collective oscillations, and the short-range interactions between individual particles (ibid.).

The 'effective separation' that the authors speak of shows itself in this: whereas in H_0 the term H_{inter} contains a mixture of particle and field coordinates, in $H_{(1)}$ the term H_{osc} contains only collective field coordinates, and $H_{part\ int}$ contains only collective particle coordinates. Bohm and Pines gather together the approximations that allow them to write the Hamiltonian in this way under the title 'The Collective Approximation' (BP I: 628a–b); I will say more about them in my comments (section 2.2.5). Among them, the 'Random Phase Approximation' plays a particularly important role in the authors' project. The 'dispersion relation' they refer to relates the frequency of the oscillation to its wavelength. This relation must hold if sustained oscillations are to occur (BP I: 625b).))

---

[11] The field coordinates appear when the vector potential of the electromagnetic field is expanded as a Fourier series. For more on Fourier series in this context, see n. 13 and the synopses of BP II and BP III.

[12] An elementary example of a move to a collective coordinate is the use of the vector $\mathbf{X}$ to specify the position of the centre of gravity of a system of masses, $m_1, m_2, \ldots, m_n$. If the positions of the masses are given by $\mathbf{x}_1, \mathbf{x}_2, \ldots, \mathbf{x}_n$, then $\mathbf{X} = \Sigma_i\ m_i\mathbf{x}_i / \Sigma_i\ m_i$.

In both the classical and the quantum mechanical accounts, the authors claim (BP I: 634b), the move to collective variables shows that, within an electron gas:

[T]he effects of magnetic interactions are divided naturally into the two components discussed earlier:

The long range part [given by H_{osc}] responsible for the long range organized behaviour of the electrons, leading to modified transverse field oscillations …

The short-range part, … given by $H_{part\ int}$, which does not contribute to the organized behaviour, and represents the residual particle-interaction after the organized behaviour of the system has been taken into account.

BP I is essentially a preamble to the papers that follow. The magnetic interactions that produce transverse oscillations are many orders of magnitude weaker than the Coulomb interactions that produce longitudinal plasma waves, and, consequently, 'are not usually of great physical import' (BP I: 627a). The authors investigate them 'to illustrate clearly the techniques and approximations involved in our methods', since 'the canonical treatment of the transverse field is more straightforward than that of the longitudinal field' (ibid.).

2.1.3. BP II: Collective *vs* Individual Particle Aspects of the Interaction

BP II and BP III both give theoretical treatments of longitudinal oscillations. A classical treatment in BP II is followed by a quantum mechanical treatment in BP III. In both these papers, as in BP I, the authors' chief concern is the relation between the individual and the collective aspects of electronic behaviour in plasmas.

In BP II, Bohm and Pines analyse this behaviour in terms of the variation of the electron density (the number of electrons per unit volume) within the plasma. Because of the forces of repulsion between electrons, these variations act like variations of pressure in air, and can be transmitted through the electron gas as plasma oscillations, like sound waves. To analyse the resulting variations in electron density no transformations of the kind used in BP I are required, since the electron density is already a collective quantity. The authors work with Fourier components ρ_k of this density.[13]

---

[13] A simple example of a Fourier decomposition occurs in the analysis of musical tones; the note from a musical instrument can be broken down into a fundamental, together with a set

The component ρ_0 represents the mean electron density of the gas, but the other components ρ_k, with $k > 0$, represent density fluctuations of different wavelengths.

((Bohm and Pines use elementary electrostatics, together with the random phase approximation, to obtain an expression for $d^2\rho_k/dt^2$ that can be divided into two parts (BP II: 340b). One part shows the contribution of the interactions between electrons, and the other the contribution of their random thermal motions.

The authors show that, if thermal motions are ignored, sustained oscillations of frequency ω_P can occur; ω_P is the so-called *plasma frequency*.[14] If thermal motions are taken into account, the frequency ω of oscillation is no longer independent of the wave number k, but is given by the approximate dispersion relation, $\omega^2 = \omega_P^2 + k^2 <v_i^2>$ (BP II: 342a). Here $<v_i^2>$ is the mean square velocity of the electrons. In the expression for ω^2, as in that for $d^2\rho_k/dt^2$, the second term represents the disruptive effect of random thermal motions. For small values of k (i.e. for long fluctuation wavelengths), the term $k^2 <v_i^2>$ is very small (as is the corresponding term in the expression for $d^2\rho_k/dt^2$). Thus, for small k, the first term in that expression predominates, and the electron gas displays its collective aspect. Conversely, for high values of k and short wavelengths, the second term predominates, and the system can be regarded as a collection of free particles. In the general case, both aspects are involved.))

Bohm and Pines show (BP II: 342b–3a) that in the general case each component ρ_k of the electron density can be expressed as the sum of two parts:

$$\rho_k = \alpha_k q_k + \eta_k$$

of overtones. More abstractly, let fx be any continuous function, such that $fx_1 = fx_2$ and fx has only a finite number of maxima and minima in the interval $[x_1, x_2]$. Then fx can be represented in that interval as the sum of a set of sine waves: $fx = \Sigma_k a_k e^{ikx}$. The index k specifying each component runs over the integers, 0, 1, 2, ... It is the *wave number* of the component of the function. That is, the number of complete wavelengths of that component in the interval $[x_1, x_2]$. Thus, k is inversely proportional to the wavelength λ; the greater the wave number the shorter the wavelength and vice versa. Messiah 1956: 471–8 gives a useful mathematical account of Fourier transformations.

[14] The plasma frequency is given by $\omega_P^2 = 4\pi ne^2/m$, where e and m are the electron's charge and mass, respectively, and n is the electron density. An elementary classical derivation of this equation is given by Raimes 1961: 283–4.

Here α_k is a constant, q_k is a collective coordinate that oscillates harmonically, and η_k describes a fluctuation associated with the random thermal motion of the electrons. Two general conclusions are drawn. (i) Analysis of η_k shows that each individual electron is surrounded by a region from which the other, negatively charged electrons are displaced by electrostatic repulsion. That region, however, like the entire volume occupied by the electron gas, is assumed to contain a uniformly distributed positive charge. The result is that the Coulomb force due to the individual electron is screened off outside a radius of the order of λ_D, the so-called 'Debye length'.[15] In a metal, this region is about 10^{-8} cm in diameter; it moves with the electron, and can itself be regarded as a free particle.[16] (ii) It turns out (BP II: 343a) that there is a critical value k_D of k such that, for $k \geq k_D$, $\rho_k = \eta_k$ (i.e. there are effectively no collective coordinates q_k for k greater than k_D). Since, by definition, $k_D = 1/\lambda_D$, the physical import of this is that there are no plasma oscillations of wavelengths less than λ_D. As in BP I, the use of collective coordinates allows Bohm and Pines to predict that long-range interactions in the electron gas give rise to plasma oscillations (in this case, longitudinal oscillations), and that at short ranges the normal electrostatic forces between electrons are largely screened off.

The authors go on to show how oscillations may be produced by a high-speed electron moving through the electron gas (BP II: 344b–7a). A 'correspondence principle argument', which I will comment on in sections 2.2.3 and 2.2.6, is then used (BP II: 347a) to explain the experimentally obtained phenomenon that, when high-energy electrons pass through thin metallic foils of the same element, they suffer losses of energy that are all multiples of a specific value. These are the results cited by Feynman; they were obtained by G. Ruthemann and by W. Lang independently in the early 1940s (see BP II: 339a and Pines 1987: 77).

2.1.4. BP III: Coulomb Interactions in a Degenerate Electron Gas

BP III is the longest and most intricately argued paper of the four. The quantum mechanical analysis it presents uses the theoretical strategy deployed

---

[15] The *Debye length* λ_D was first introduced in connection with screening processes in highly ionized electrolytes (see Feynman, Leighton, and Sands 1963–5: ii. 7–9). It is the thickness of the ion sheath that surrounds a large charged particle in an electrolyte BP II: 341b (n).

[16] A similar result appears in BP III. Here Bohm and Pines anticipate what became standard practice in the 1950s and 1960s, whereby a particle together with its interactions with its immediate environment was treated as an elementary system, and was referred to as a *quasi-particle*. I return to this topic in section 2.3.1

in the latter part of BP I alongside the physical insights achieved in BP II. As in BP I, a Hamiltonian for the system electrons-plus-field in terms of individual particle coordinates is transformed to one in terms of collective coordinates. In this case, however, the collective coordinates are those appropriate for describing longitudinal, rather than transverse, oscillations, and the procedure is far from straightforward. A sequence of five modifications, which include coordinate transformations, algebraic manipulations, and a variety of approximations (some relying on non-trivial assumptions) takes the reader from the initial Hamiltonian (which I will refer to as 'H^1') to the final Hamilton H_{new}.[17]

H^1 is itself a standard Hamiltonian for 'an aggregate of electrons embedded in a background of uniform positive charge' (BP III: 610a). It contains three terms, one for the kinetic energy of the electrons, another for the energy due to the pairwise electrostatic interactions between them, and the third for the sum of the individual self-energies of the electrons. The last two are slightly adjusted to take into account the background of positive charge. The first four modifications of the Hamiltonian are effected in section II of the paper. By means of them, Bohm and Pines introduce a set of 'field variables', and analyse their 'approximate oscillatory behavior' (BP III: 611b). The final modification, made in section III, enables them 'to carry out the canonical transformation to the pure collective coordinates' (ibid.).

((In the first modification, H^1 is transformed into another Hamiltonian H^2, expressed in terms, not just of individual particle coordinates, but also of the longitudinal vector potential $\mathbf{A(x)}$ and the electric field intensity $\mathbf{E(x)}$ of the electromagnetic field within the plasma. The transformation does not affect the first and third terms of H^1. The difference between H^1 and H^2 is that the energy that in H^1 was attributed to Coulomb interactions between electrons is now represented as an energy of interaction between individual electrons and the electromagnetic field engendered by the electron gas as a whole. Like the density ρ in BP II, $\mathbf{A(x)}$ and $\mathbf{E(x)}$ are both written as

---

[17] The Hamiltonians that I refer to as $H^1, H^2, \ldots, H_{new}$ appear in BP III as follows: H^1 is the Hamiltonian (610b); H^2, the Hamiltonian H (612a); H^3, the Hamiltonian H (612a); H^4 is a Hamiltonian whose last four terms appear explicitly (612b), and whose first term, H_{part}, is the sum of the first and last terms of H^3, where they represent the kinetic and self energies of the electrons, respectively; H^5 is the Hamilton H (616b); H_{new} appears on p. 618b.

Fourier series (see n. 13), and the coefficients of the components p_k and q_k of the two series (both indexed by k) serve as field coordinates of the plasma, as in BP I. The authors claim (BP III: 611a) that H^2 will 'lead to the correct equation of motion', when supplemented by a set of 'subsidiary conditions' on the allowable states Φ of the system. Each member of the set has the form:

$$\Omega_k \Phi = 0, \text{ where } \Omega_k = p_{-k} - i(4\pi c^2/k^2)^{1/2} \rho_k$$

Here, ρ is the electron density, and each condition corresponds to a component of the Fourier decompositions of p and ρ.

These subsidiary conditions also serve other functions. They ensure conformity with Maxwell's equations, and also achieve a more technical purpose, that of reducing the number of degrees of freedom allowed by the transformed description (H^2 plus the conditions) to that allowed by H^1.[18] In addition, as the expression for Ω_k makes clear,

[They] introduce in a simple way a relationship between the Fourier components of the electronic density ρ_k and a set of field variables p_k. ... [T]here is in consequence a very close parallel between the behaviour of the ρ_k, as analysed in [BP] II: and the behaviour of our field coordinates.   (BP III: 611b)

Bohm and Pines extend the parallel further. They anticipate that, just as in the classical theory of plasmas developed in BP II, there is a minimum wavelength λ_c of organized oscillations, and a corresponding maximum wave number k_c, so in the quantum theory a similar (but not identical) cut-off exists (BP III: 611b). Accordingly, they use a modified version of the transformation that yielded H^2 to obtain an operator H^3. The effect of this modification is to eliminate field coordinates with wave vector k greater than some critical wave vector k_c, and so confine attention to k-values between $k = 0$ and $k = k_c$. Of H^3, Bohm and Pines remark (BP III: 616a):

There is a close resemblance between [this] Hamiltonian, which describes a collection of electrons interacting via longitudinal fields, and the Hamiltonian $[H_0]$ we considered in BP I, which described a collection of electrons interacting via the transverse electromagnetic fields. ... [O]ur desired canonical transformation is just the longitudinal analogue of that used in BP I to treat the organized aspects of the transverse magnetic interactions in an electron gas.

---

[18] The fact that the transformation allowed the system too many degrees of freedom gave Bohm and Pines considerable trouble; see Pines 1987: 75.

Algebraic manipulation shows that H^4 can be represented schematically as:

$$H^4 = H_{\text{part}} + H_1 + H_{\text{osc}} + H_{\text{s.r.}} + U$$

Here H_{part} represents the energy of particles (electrons) due to their motion and self-energy; H_1 'represents a simple interaction between electrons and the collective fields'; H_{osc} is 'the Hamiltonian appropriate to a set of harmonic oscillators, representing collective fields'; and $H_{\text{s.r.}}$ 'represents the short range part of the Coulomb interaction between the electrons [i.e., the effective Coulomb interaction once the effects of screening have been taken into account]' (BP III: 612b). The authors use (and justify) the random phase approximation to show that the remaining term U can be disregarded (BP III: 613b–14b), and then rewrite H_{osc}, dubbing the result 'H_{field}'.[19] In this way they arrive at the Hamiltonian H^5:

$$H^5 = H_{\text{part}} + H_1 + H_{\text{field}} + H_{\text{s.r.}}$$

The 'subsidiary conditions' on the allowable states Φ of the system now appear as a set of k_c conditions:

$$\Omega_k \Phi = 0 (k < k_c)$$

and a new expression for the operators Ω_k is given (BP III: 616b).

Following the strategy of BP I, Bohm and Pines apply a canonical transformation to H^5 and to the Ω_k operators. The transformation is designed to 'eliminate H_1 to first order' by distributing most of its effects among the (transformed versions of) other terms of H^5, and so redescribe the system in terms of 'pure collective coordinates'.[20] That is to say, no term of

---

[19] H_{field} is obtained in BP III: 616a by replacing each occurrence of q_k in H_{osc} by the expression $\hbar\omega^{1/2}a_k - a_k{}^*$. The reason for making this move will be made clear in section 2.2.7 of this essay. q_k represents a component of the Fourier expansion for the longitudinal vector potential of the electromagnetic field $\mathbf{A}x$, and a_k and $a_k{}^*$ are, respectively, the creation and annihilation operators for the longitudinal photon field (BP III: 616a). The equation $q_k = \hbar/2\omega^{1/2}a_k - a_k{}^*$ is an identity in the operator algebra used in quantum mechanics. For an introduction to creation and annihilation operators, see Messiah 1958: 438–9 and 963–6. For the mathematical definition of a transformation in this context, see the examples 3, 6, and 7 in section 2.2.3 of this essay.

[20] What is left of H_1 is then discarded, and the resulting Hamiltonian symbolized by '$H_{\text{new}}{}^0$' to mark the fact that it is a lowest-order approximation. In section III of BP III, Bohm and Pines then show that this residue of the transformed version of H_1 may be neglected, and so the superscript '0' is dropped.

the transformed Hamiltonian contains both particle and field operators. The modifications have done their work.))

The final Hamiltonian, which Bohm and Pines christen 'H_{new}', differs markedly from the Hamiltonian H^1 from which they started. It is expressed schematically as:

$$H_{new} = H_{electron} + H_{coll} + H_{res\ part}$$

Like $H_{(0)}$ in BP I, this Hamiltonian contains just three parts. $H_{electron}$ contains terms referring only to individual electrons; field coordinates appear only in H_{coll}, which describes independent longitudinal oscillations of the field; and $H_{res\ part}$ represents an extremely weak 'residual' electron–electron interaction; at short range the electrons are effectively screened off from each other. Thus, the quantum mechanical treatment of the plasma has replicated the conclusions of the classical treatment in BP II.

((The oscillations described by H_{coll} are independent in the sense that, under the canonical transformation (i) the subsidiary conditions that guarantee conformity with Maxwell's equations no longer relate field and particle variables and (ii) H_1, which represented field–particle interactions in H^3, H^4, and H^5, has now disappeared; effectively, the final transformation has distributed it between two terms of the Hamiltonian.[21] Part of it reappears as $H_{res\ part}$; the electron–electron interaction this operator describes is negligible in comparison with the short-range interaction described by $H_{s.r.}$ in H^4. The other part has been absorbed into $H_{electron}$, where it appears as an increase in the 'effective mass' of the electron. Bohm and Pines interpret this (BP III: 620a–b) as 'an inertial effect resulting from the fact that these electrons carry a cloud [of collective oscillations] along with them' (see n. 16 above).))

In this way the programme announced in BP I has been carried out. Bohm and Pines have demonstrated 'explicitly in Hamiltonian form the effective separation between long range collective interactions, described here in terms

[21] Note that in BP III the expressions H_{part} and $H_{s.r.}$, denoted components of H^4. Since H_{new} is a transformed version of H^4, it would have been better to denote the corresponding terms in H_{new} by H^C_{part} and $H^C_{s.r.}$, in order to reflect the fact that pure collective coordinates are being used. Instead, in P IV 627a, they appear, misleadingly, as H_{part} and $H_{s.r.}$.

of organized oscillations, and the short range interactions between individual particles' (BP I: 626b).

2.1.5. P IV: Electron Interaction in Metals

In P IV, the last paper in the series, Pines applies the quantum mechanical account of plasmas developed in BP III to the behaviour of the valence electrons in metals (otherwise known as 'conduction electrons'). He begins by drawing attention to the assumptions this involves (P IV: 626a). In BP III, the system described by the initial Hamiltonian H^1 consisted of cloud of electrons moving against a uniform background of positive charge. In a metal, however, the positive charge consists of positive ions localized on the nodes of the crystal lattice; in addition, this lattice undergoes vibrations. Thus, when the results of BP III are carried over into P IV, the first assumption made is that the periodicity and the density fluctuations of the positive charge in a metal can be ignored. The second is that 'the only interactions for the conduction electrons in a metal are those with the other conduction electrons' (P IV: 626a).

((Pines acknowledges that 'if, for instance, the exchange interaction with the core electrons is large, the collective description may well become inapplicable' (ibid.), since the validity of the collective description requires that:

[T]he mean collision time for electron collisions ... should be large compared with a period of oscillation. This follows from the fact that the effect of disruptive collisions is to cause damping of the collective oscillations. [*See section 2.1.4 of this essay.*] If ... the damping is large, ... the whole concept of collective oscillations loses its significance in a description of electron interaction.))

An important goal throughout P IV is to show that, despite the simplifications and idealizations involved, the theoretical results of BP III hold for the electron interactions in (at least some) metals.[22] The first item on Pines' agenda is to show how the collective account can explain why, for many purposes, the independent electron model (described here in section 2.1.1) worked as well as it did. His argument (P IV: 627a–b) is very simple. He reminds the reader that two important mathematical results have been obtained in BP

---

[22] Pines remarks (P IV: 626): 'This assumption should be quite a good one for the alkali metals ... and we may expect it to apply generally for any metallic phenomena in which the periodicity of the lattice plays no important role.'

III: an expression for the Hamiltonian H_{new} for the system, and the subsidiary conditions Ω_k on its states. The collective description of the electron gas that H_{new} provided included a term H_{coll} which summarized, so to speak, the effects of the long-range Coulomb interactions between electrons. They were 'effectively redescribed in terms of the collective oscillations of the system as a whole' (ibid.). But, given the dispersion relation for these oscillations, which relates their frequency to their wavelength, 'It may easily be shown that the energy of a quantum of effective oscillations is so high … that these will not normally be excited in metals at ordinary temperatures, and hence may not be expected to play an important role in metals under ordinary conditions' (ibid.). Pines goes on:

The remainder of our Hamiltonian corresponds to a collection of individual electrons interacting via a comparatively weak short-range force, $H_{s.r.}$. These electrons differ from the usual 'free' electrons in that they possess a slightly larger effective mass [*see section 2.1.4 of this essay*], and their wave functions are subject to a set of [subsidiary] conditions. However, both of these changes are unimportant qualitatively (and in some cases quantitatively). Furthermore, since the effective electron–electron interaction is so greatly reduced in our collective description, we should expect that it is quite a good approximation to neglect it for many applications. Thus, we are led directly to the independent electron model for a metal.

Like the arguments of BP III, this argument arrives at a theoretical conclusion about an electron gas through an examination of its theoretical description, the Hamiltonian H for the system.

((The Hamiltonian Pines uses is a modified version of the Hamiltonian H_{new} obtained in BP III. Whereas in that paper Bohm and Pines wrote:

$$H_{new} = H_{electron} + H_{coll} + H_{res\ part}$$

in P IV Pines rewrites $H_{electron}$ as $H_{part} + H_{s.r.}$, and neglects $H_{res\ part}$ since 'it will produce negligible effects compared with $H_{s.r.}$, and this latter term is small' (P IV: 627a). The resulting Hamiltonian is:

$$H = H_{part} + H_{coll} + H_{s.r.}))$$

In the remainder of P IV, Pines examines the quantitative results that this Hamiltonian yields. In section II, he uses it to calculate the energy ε_0 of the system in its ground state, and compares the result with those reached by other approaches. In section III, he points to a problem encountered by the independent electron approach, and suggests how it could be accounted for by the Boham-Pines (BP) theory. In section IV, he shows how that theory can also explain the behaviour of high-energy particles as they pass through metal foils.

((Pines begins section II by pointing out that, given the Hamiltonian for the system, the direct way to obtain ε_0 would be to solve the time-independent Schrödinger equation:

$$H_{\text{new}}\psi_0 = \varepsilon_0\psi_0$$

in which ε_0 appears as the lowest eigenvalue corresponding to the eigenfunction ψ_0. Instead, for ease of calculation, he discards the smallest term $H_{\text{s.r.}}$ of H_{new}, and works with a wave function ψ_0, which is both an exact eigenfunction of the resulting Hamiltonian ($H_{\text{part}} + H_{\text{coll}}$), and an approximate eigenfunction of H_{new}.[23] He argues (P IV: 630a–b) that, since the contribution of the neglected term $H_{\text{s.r.}}$ is small, it can be treated later as a small perturbation in H_{new}, giving rise to a small short-range 'correlation energy' $\varepsilon_{\text{corr.}}$.[24] For details of the calculation Pines refers the reader to a forthcoming paper.

Here the BP theory and the independent electron model part company. When ε_0 is compared with the energy E as calculated on (one version of) the independent electron model, the two values differ by a small amount which Pines calls a 'correlation energy'. This can be broken into two parts, corresponding to a long-range and a short-range interaction between electrons. Symbolically,

$$\varepsilon_0 - E = \varepsilon_{\text{corr.}} = \varepsilon_{\text{corr}}^{\text{l.r.}} + \varepsilon_{\text{corr}}^{\text{s.r.}}$$

In section III, Pines shows that, when this 'correlation energy' is taken into consideration, the independent electron model faces a problem. When all electron–electron interactions are neglected, and the energy of the electron gas is taken to be the Fermi energy E_F,[25] many of the results obtained using

---

[23] Note that, even though he works with an approximate eigenfunction, Pines still denotes it by 'ψ_0'.

[24] I say more about perturbation theory in example 4 and n. 44, which accompanies it in the discussion of the use of theory in section 2.2.3, and also in section 2.2.7.

[25] According to the Pauli exclusion principle, no more than one electron can occupy a given state. To each state there corresponds a certain energy level. In a gas of n electrons at absolute zero the first n levels will be filled, and the gas will have a corresponding energy. This is called the *Fermi*

the model agree quite well with experiment. Quantum mechanics, however, decrees that one should allow for an additional 'exchange energy' E_{exch},[26] and when the energy is 'corrected' from E_F to $E_F + E_{exch}$, the agreement with experiment, far from being improved, is rendered worse than before. Pines illustrates this with two examples. The first (P IV: 631a) concerns the specific heat of an electron gas. According to the 'corrected' version of the independent electron model, this quantity will vary as $T/\ln T$ (where T is the absolute temperature).[27] Experiment, however, shows a linear dependence on T. The second example (P IV: 632a–b) concerns the magnetic properties of an electron gas. According to the 'corrected' account, the electron gas in certain metals—cesium, for example—would display ferromagnetic behaviour (i.e. the spins of the electrons could become aligned). But, no such behaviour is observed. Neither anomaly arises with the simpler version of the independent electron model.

In contrast, on the BP model, the long-range Coulomb interactions that lead to a large exchange energy E_{exch} are replaced by effective screened short-range Coulomb interactions (see section 2.1.4), with a corresponding reduction in energy contributions. The net result is that the effect of exchange energy on the specific heat of the electron gas is comparatively slight, and the model never displays ferromagnetic behaviour (P IV: 623a–b).))

In the final section of P IV, Pines returns to a phenomenon discussed in BP II: the excitation of plasma oscillations by high-energy charged particles passing through a metal. To describe the motion of the particle and its interaction with the electron gas, he adds two terms to the Hamiltonian H^5 of BP III; in order to rewrite these terms in collective coordinates, he then applies to them the same canonical transformation that took H^5 into H_{new} in BP III; lastly, he uses the random phase approximation, the dispersion relation, and the subsidiary conditions of BP III to obtain the three-term Hamiltonian H_{add} (P IV: 633a–b). The first term describes 'a short-range screened Coulomb interaction between the charged particle and the individual electrons in the electron gas' (P IV: 633a); the second 'the interaction between the charged

energy. Because the energy levels are close together, for modest values of T the energy will not change much, since every electron that jumps to a level just above the nth level vacates a level just below it.

[26] If two similar quantum systems—in this case, two electrons—are close together, quantum mechanics tells us that their combined state must be such that it would not be altered if the two systems exchanged their individual states. The effect adds a small energy E_{exch} to the energy of the composite system.

[27] Here Pines draws on theoretical work by James Bardeen and E. P. Wohlfarth (see P IV: 631a).

particle and the collective oscillations of the system' (P IV: 633b); the third may be neglected. Analysis of this second term (P IV: 633b–4a) shows that the interaction generates forced oscillations in the collective field, which at a certain frequency ω_P will produce resonant oscillations in the electron gas. By the same argument as was used in BP II: since the energy associated with an oscillation of frequency ω_P is $\hbar\omega_P$ (where $\hbar$ is Planck's constant), the total energy loss suffered by a high-energy particle in exciting such oscillations should be some multiple of $\hbar\omega_P$. In this way, energy is transferred from the particle to these oscillations in discrete quanta. The quantum of energy loss sustained by the particle in each transfer can be calculated, and the results agree well with the experimental findings of Ruthemann and Lang mentioned earlier.

Finally, Pines uses the first term of H_{add} to obtain an expression for the 'stopping power' of a metal, the loss of energy by a charged particle per unit length of the distance it travels through the metal (P IV: 635a–b). He compares his results with those predicted by Aarne Bohr (1948) and H. A. Kramers (1947), each using a different theoretical approach, and with those obtained experimentally for lithium and beryllium by Bakker and Segré (635b).[28] All four sets of results coincide, within the ranges of error of their respective theoretical and experimental practices.

2.2. OBSERVATIONS ON THE BOHM–PINES QUARTET

2.2.1. Introduction

The physicist John Ziman offers this description of 'the intellectual strategy of a typical paper in theoretical physics':

A model is set up, its theoretical properties are deduced, and experimental phenomena are thereby explained, without detailed reference to, or criticism of, alternative hypotheses. Serious objections must be fairly stated; but the aim is to demonstrate the potentialities of the theory, positively and creatively, 'as if it were true'.   (1978: 3–4)

Individually and collectively, the four Bohm–Pines papers all conform to this description; indeed it might have been written with them in mind.[29] My

---

[28] Kramers used 'a macroscopic description in which the electrons were treated as a continuum characterized by an effective dielectric constant'. We will meet it again in Part 2.3.

[29] In point of fact, Ziman had read these papers carefully (see Ziman 1960: 161–8).

comments on them are grouped under six headings: the use of models, the use of theory, the modes of description offered, the use of approximations, the connection with experiment, and the nature of deduction as it appears in the quartet. My aim is to provide descriptions, rather than evaluations, of the theoretical practices in which Bohm and Pines engaged. But, before I embark on this project, a few remarks about Ziman's capsule summary are called for. Like the papers in theoretical physics it describes, the summary itself takes a lot for granted. A typical paper may well begin by setting up a model, but this is not done in a theoretical vacuum. Physicists inherit the achievements of their predecessors, and yesterday's models become today's physical systems, waiting to be modelled in their turn. Bohm and Pines, for example, take for granted, first, the account of a plasma as an electron gas that contains enough positively charged ions to neutralize the electrons' charge, and secondly, the assumption that a metal is a special case of such a system, whose distinguishing feature is the systematic spacing of the ions. Both assumptions are implicit in the references Pines makes to 'the ionic field', to 'laboratory plasmas', and (in a particularly telling phrase) to 'the actual plasma' in the quotations with which the next section begins.

2.2.2. The Use of Models

In the quartet, Bohm and Pines reserve the word 'model' for the independent electron model. Yet their own approach also relies on a highly simplified model of a metal, one which enabled them to treat it like a gas plasma. By 1955, Pines describes it in just those terms (1955: 371): '[W]e shall adopt a simplified model for a metal in which we replace the effect of the ionic field by a uniform background of positive charge.' And, whereas in 1951 this theoretical move is relegated to a footnote (BP I: n. 1), Pines subsequently accords it much more importance. He writes:[30]

In any approach to understanding the behaviour of complex systems, the theorist must begin by choosing a simple, yet realistic model for the behaviour of the system

---

[30] This paragraph was written in the mid-1960s. It formed part of an unfinished chapter of a book. Pines tells us (1987: 84): 'I had hoped to describe for the layman how theoretical physicists actually work, how a unified theory develops out of the interplay between physical concepts and mathematical reasoning, between theory and experiment, and between theorist and theorist. I found the task difficult, if not impossible, and put the book aside.' The example he used in this unpublished chapter was the work on electron interactions in metals that he and Bohm had done together. Fittingly enough, twenty years after it was written, this chapter formed the first half of the article Pines contributed to a *Festschrift* in Bohm's honour (Hiley and Peat 1987).

in which he is interested. Two models are commonly taken to represent the behaviour of plasmas. In the first, the plasma is assumed to be a fully ionized gas; in other words as being made up of electrons and positive ions of a single atomic species. The model is realistic for experimental situations in which the neutral atoms and the impurity ions, present in all laboratory plasmas, play a negligible role. The second model is still simpler; in it the discrete nature of the positive ions is neglected altogether. The plasma is then regarded as a collection of electrons moving in a background of uniform positive charge. Such a model can obviously only teach us about electronic behaviour in plasmas. It may be expected to account for experiments conducted in circumstances such that the electrons do not distinguish between the model, in which they interact with the uniform charge, and the actual plasma, in which they interact with positive ions. We adopt it in what follows as a model for the electronic behaviour of both classical plasmas and the quantum plasma formed by electrons in solids. (1987: 68)

This paragraph is very revealing. Pines tells us that:

1) The theorist's task involves choosing a model.

2) In dealing with a complex system, that is the only option available.

3) The model can help us to understand the behaviour of a system; in other words, it can provide explanations of that behaviour (a claim echoed by Ziman).

4) The model is to be 'simple, yet realistic'.

5) The model will involve some simplification; in the first model Pines describes, impurities and neutral atoms are to be disregarded.[31]

6) A model may misrepresent aspects of the system; a regular array of positive ions may be represented as a background of uniformly distributed positive charge.

7) More than one way of modelling may be used.

We may also observe that:

8) The components of the models—electrons, positive ions—would be described in standard philosophical parlance as 'theoretical entities'.

9) The two models Pines describes are at odds with one another.

10) The model he adopts involves a greater degree of misrepresentation than the other.

[31] What I have called 'simplification', some philosophers refer to as 'abstraction'. See, for example, the footnote at Morgan and Morrison 1999: 38.

In the paragraph I quoted, Pines deals with a particular physical system and the ways in which it can be modelled. In contrast, among the aspects of modelling I have listed, the first eight are very general, and analogues of aspects 9 and 10 appear frequently in theoretical practice. Some amplification of these points is called for.[32]

Note first that, in addition to the simplifications already mentioned, there are interactions that some models cannot accommodate. Because they ignore the fact that ions form a regular lattice, they cannot take into account interactions between electrons and lattice vibrations (which are instrumental in bringing about superconductivity).[33] In addition, the only electron–electron interactions considered are between conduction electrons; interactions between conduction electrons and core electrons (those immediately surrounding the positive nuclei of the ions) are assumed to have no importance.[34]

Despite these simplifications, and items 5–10 above, Pines tells us that a model should be 'realistic' [4], and claims that, for certain experimental situations, the model he uses meets that criterion. A contrasting view is expressed by Conyers Herring. In commenting on a theoretical account of the surface energy of a metal (Ewald and Juretschke 1952) he observes (117):

It is to be emphasized that the wave mechanical calculation of the surface energy given in the paper applies not to a real metal, but to a a fictitious metal ... The fictitious metal consists, as has been explained, of a medium with a uniform distribution of positive charge—we may call it a 'positive jelly'—and a compensating number of electrons. This metal ... we may call 'jellium' to distinguish it from real metals such as sodium.

The same model that Pines describes as 'realistic' is here described by Herring as 'fictitious'. There is no direct contradiction between these descriptions; works of fiction can be realistic. And, while the jellium model is not realistic in quite this sense, jellium itself is endowed by Herring with physical properties: 'A rough calculation ... has indicated that jellium of an electron density equal to that of sodium should have a binding energy only about two thirds that of sodium' (ibid.).

---

[32] In the remainder of this section, a numeral placed in parentheses (e.g., '[9]') draws attention to the specific point being discussed.

[33] I return to this interaction in Section 2.3.1 of this essay.

[34] Pines himself draws attention to these simplifications at the start of P IV: 626a.

By using the term 'realistic', Pines marks a distinction between two types of model: the models he himself works with and *analogue* models like the liquid-drop model of the nucleus proposed by Niels Bohr in the late 1930s. An analogue model relies on a correspondence between the behaviours of two otherwise radically different types of physical systems (nuclei and liquid drops, for example). A 'realistic' model, on the other hand, is defined by taking a description of the physical system to be modelled, and modifying it by a process of simplification and idealization [5, 6]. These descriptions will be in a well understood vocabulary that includes terms like 'electron' and 'positive ions'. The philosopher may regard such entities as merely 'theoretical' [8], but when Pines used them they had been familiar and accepted elements of the physicist's world for forty years.

But, why are such models *necessary*, as Pines insists [1, 2]? They are needed because without them it would be impossible to apply our theories to the physical world. In the first place, given a 'complete description' of a natural system, we still need a principle by which irrelevancies can be winnowed out from salient information. Secondly, from Galileo onwards, our theories of physics have treated only ideal entities (point masses, rigid bodies, frictionless planes), items that are absent from the physical world. As we have seen, a realistic model of the kind Pines envisages is an entity defined in terms of simplifications and idealizations [5, 6]. Effectively, the model's definition allows it to act as a principle of selection, and the idealizations built into it make it amenable to treatment by our theories. A model of this kind functions as an essential intermediary between the theories of physics and the physical world.[35]

How does a model help us to understand the behaviour of a physical system [3]? The short answer is that models are things we can play with. A model's resources are gradually made available to us as we come to see how much can be deduced from its theoretical definition, and how aspects of its behaviour are interlinked. As a variety of phenomena are successfully represented by the model, it becomes progressively more familiar to us; increasingly, we come to see the physical system in terms of the model, and vice versa.[36]

[35] This point has been made by many writers, including Nancy Cartwright (1983; 1999), Ernan McMullin (1985), Ronald Giere (1985), Mauricio Suarez (1999), and Margaret Morrison and Mary Morgan (1999). Morrison and Morgan go on to emphasize that models act as mediating instruments in a great variety of ways in addition to the one I discuss here.
[36] For an extended account along these lines, see Hughes 1993.

Because a model of this kind is a defined entity, pragmatic considerations can influence what simplifications and idealizations are made [6]. The theorist has to make choices [7, 9]. An additional idealization may make his problems more tractable, but at the cost of making his predictions less accurate. Luckily these choices are not irrevocable. In making that idealization, the theorist will be opting for the model that involves the greater degree of misrepresentation [10], but he can always rescind it in the search for greater empirical adequacy. Pines did just that. Throughout the quartet he and Bohm opted for the jellium model, and treated the positive charge of metallic ions as uniformly distributed. Three years later, Pines (1956) treated the charge more realistically, as one that varied regularly within the metal, in order to make more accurate estimates of the energy loss of high-energy particles in metal foils.

2.2.3. The Use of Theory

After a model has been set up, Ziman tells us, 'its theoretical properties can be deduced'. But, as we learn from Pines, the same model can be adopted 'as a model for the electronic behaviour of both classical plasmas and the quantum plasma formed by electrons in solids'. Which is to say (i) that the behaviour of the model is assumed to be governed by a foundational theory, and (ii) that in some cases it is appropriate to use classical mechanics, and in others, quantum mechanics. We have seen this happen, first in BP I, which provided both classical and quantum mechanical treatments of magnetic interactions between electrons, and then in BP II and BP III, which provided, respectively, classical and quantum mechanical accounts of electrostatic (Coulomb) interactions. In addition, the use of a 'correspondence principle argument' in BP II can be seen as an appeal to the old (pre-1925) quantum theory (see section 2.1.3.). A wholly classical analysis is given of the energy loss suffered by a high-velocity charged particle when it excites plasma oscillations; this analysis is then supplemented by the assumption that the energy loss is quantized, and that the energy E per quantum is given by the Planck formula, $E = \hbar\omega$. Here $\hbar$ is Planck's constant, and ω is the frequency of the plasma oscillation which it excites.

Early quantum theory was notorious for its reliance on *ad hoc* procedures. To quote Max Jammer (1966: 196):

In spite of its high sounding name ... quantum theory, and especially the quantum theory of polyelectronic systems, prior to 1925 was, from the methodological point of view, a lamentable hodgepodge of hypotheses, principles, theorems, and computational recipes rather than a logical consistent theory.

The introduction in BP II of the Planck formula within an otherwise classical treatment of plasmas is a case in point. The formula functions simply as a useful tool in the theoretician's workshop. In contrast, the theories the authors use in BP III are post-1925 orthodox quantum mechanics and (occasionally) standard electromagnetic theory and quantum field theory. Each of these theories is a foundational theory, in the sense that it is undergirded by a powerful mathematical theory.[37] While not conceptually wrinkle-free,[38] none of them would normally be described as 'a lamentable hodgepodge'. None the less, in one respect, each of them resembles early quantum theory in the first quarter of the twentieth century. They, too, provide sets of ready-to-hand tools for the theoretician's use. To extend the metaphor, the difference is that all the tools from a given theory now come from the same tray of the toolkit.[39]

I will illustrate this use of theory with nine examples from the Bohm–Pines papers, the first very general, the rest specific. In all four papers foundational theory provides a template for the mathematical description of a system, and specifies how that form can be given content. Whether provided by classical physics or quantum mechanics, these descriptions are given by the Hamiltonian H for the system, which represents its total energy.[40] H may be written as the sum $H_1 + H_2 + \ldots + H_n$ of terms, each of them representing a different source of energy (kinetic energy, electrostatic potential energy, and so on). The mathematical nature of H is not the

---

[37] A paradigm example is the mathematical theory set out in John von Neumann's *Mathematical Foundations of Quantum Mechanics* [1932] 1955.

[38] Orthodox quantum mechanics was not a wholly unified theory, and no one has yet solved the 'measurement problem'.

[39] I was first introduced to the image of a theory as a set of tools by Paul Teller, in conversation. It is drawn by Nancy Cartwright, Tofic Shomar, and Mauricio Suarez in the paper 'The Tool-box of Science' (1994). The earliest use of it that I can trace is by Pierre Duhem in the series of articles in the *Revue de Philosophie* in 1904 and 1905 that later became his *The Aim and Structure of Physical Theory* (*PT*, 24). Edmund Husserl, in a discussion of geometrical praxis ([1954] 1970: 26) talks of 'limit-shapes [that] have become acquired tools that can be used habitually and can always be applied to something new'. Philosophically, the metaphor of language as a toolbox is well known from Ludwig Wittgenstein's *Philosophical Investigations*, §11 (1953: 6c). In Wittgenstein's metaphor, the tools in the toolbox of language are all individual words; in contrast, the tools provided by theoretical physics are not all of one kind, as will appear.

[40] In classical physics, an alternative mode of description, in terms of forces, may be used, but this is not available in quantum mechanics.

same in the two theories; in classical physics, H is a function, in quantum mechanics it is an operator.[41] There are, however, standard procedures for obtaining a quantum mechanical Hamiltonian from a classical one. To take a particular case, wherever a momentum p appears in a classical Hamiltonian function, one substitutes the operator $-i\hbar(\partial/\partial x)$ to obtain the corresponding quantum mechanical operator; thus, the kinetic energy term represented classically by the function $p^2/2m$ appears as the quantum mechanical operator $-(\hbar^2/2m)\partial^2/\partial x^2$.

Specific instances of the use of theoretical tools—results, strategies, and technical manoeuvres—which theory provides and whose use needs no justification, are supplied by eight examples from BP III. Here as elsewhere, Bohm and Pines are considering 'an aggregate of electrons embedded in a background of uniform positive charge' (610b).[42]

1. (610b) The authors write down the Hamiltonian H^1 for the system. It contains three terms: the first is the operator provided by the theory to express the kinetic energy of the electrons; the second and third are standard expressions for the energy due to Coulomb attractions between electrons, and for their self-energy; each of them is slightly modified to take into account the uniform background of positive charge, and the second term is expressed as a Fourier series (see n. 13).

2. (610b–11a) When this Hamiltonian is rewritten in terms of the longitudinal vector potential $\mathbf{A}(\mathbf{x})$ and the electric field intensity $\mathbf{E}(\mathbf{x})$, each of these quantities is expressed as a Fourier series involving one of the field coordinates q_k or p_k (for the position and the momentum associated with the field). Both of these series are supplied by electromagnetic theory.

3. (611a–b) To show that the resulting Hamiltonian H^2 is equivalent to H^1, Bohm and Pines use the method quantum mechanics prescribes:

[41] In classical physics, the Hamiltonian H of a system is a function $H: \Omega \to \mathbf{R}$, where Ω is the set of states of the system. For any state $\omega \in \Omega$, the number $H\omega$ is the total energy of the system in that state. In quantum mechanics, the Hamiltonian H of a system is an operator $H: \Psi \to \Psi$, where Ψ is the set of states of the system. A state in quantum mechanics is represented by a wave function ψ. For any state $\psi \in \Psi$, there is a state ψ_H not necessarily distinct from ψ such that $H\psi = \psi_H$. If ψ is an *eigenfunction* of H, then there is a real number E such that $H\psi = E\psi$ which is to say, E is an *eigenvalue* of H, and E is the energy of the system in state ψ, as in the synopsis of P IV. A fuller comparison of classical physics and quantum mechanics is given in Hughes 1989: chap. 2. Jordan 1968 provides a comprehensive yet concise treatment of the operators used in quantum mechanics.

[42] All page references in this subsection are to BP III. With one exception, the synopsis of BP III in section 2.1.4 shows where each of the tools was used.

they display a unitary operator S, such that $H^2 = SH^1S^{-1}$.[43] As they note (611a n.), this operator is supplied by Gregor Wentzel's textbook, *Quantum Theory of Wave Fields*.

4. (614b) The 'perturbation theory' of quantum mechanics is called on to estimate what corrections would have to be applied to compensate for the neglect of the terms U and H_1 in H^4.[44]

5. (616a) As I noted in example 2, at an early stage of BP III, Bohm and Pines introduce field coordinates q_k and p_k. Now, 'in order to point up the similarity [between the transformations used in BP III and those used in BP I] and to simplify the commutator calculus', they help themselves to the fact that in quantum mechanics these operators can be expressed in terms of creation and annihilation operators a_k and $a_k{}^*$ (see n. 19).

6. (616b–17a) The last of the transformations performed on the Hamiltonian in BP III takes H^5 into H_{new}. Like the transformation of H^1 into H^2 mentioned in example 3, this transformation is performed by a unitary operator. The authors' goal is to find a unitary operator U such that $U^{-1}H^5U = H_{new}$. A basic theorem of the algebra of operators is that any unitary operator U is expressible as an exponential function of another (non-unitary) operator S, the so-called *generator* of U; we may write $U = \exp(iS/\hbar)$.[45] Hence, Bohm and Pines set out to obtain U by finding a suitable generator S.

This was a perfectly orthodox strategy to employ; in a footnote (616b), the authors cite the second edition of P. A. M. Dirac's classic text, *The Principles*

[43] For more on unitary operators, see Jordan 1969: 18–22. S is a *unitary operator* if there is an operator S^{-1} such that $S\,S^{-1} = I = S^{-1}\,S$, where I is the identity operator: for all ψ, $I\psi = \psi$. S^{-1} is the *inverse* of S; for any wave function ψ_1, if $S\psi_1 = \psi_2$, then $S^{-1}\psi_2 = \psi_1$.

[44] The use of perturbation techniques in physics goes back to Isaac Newton. In quantum mechanics, perturbation theory is invoked when a term that makes a very small contribution to the Hamiltonian is neglected to simplify calculations. Its function is to estimate the correction that would have to be applied to the result obtained in order to allow for the effect of that term. A concise account of the principles of first-and second-order perturbation theory in quantum mechanics is given by Cassels (1970: 780–80); for more details, see Messiah 1958: chaps. 16–17. To estimate corrections for U and H_1 in H^3, Bohm and Pines use second-order perturbation theory.

[45] See Jordan 1968: 52. Also, two remarks on notation: (a) In example 3, I used the letter 'S' for the unitary operator that transforms H^1 into H^2, but in this example I use it for the *generator* of the unitary operator that transforms H^5 into H_{new}, rather than the unitary operator itself; (b) I write 'U' for the second unitary operator, even though the same letter is used elsewhere to denote a term in the Hamiltonian H^3; I have no one to blame but myself. Within this example, 'U' always denotes a unitary operator and 'S' its generator. See [7] in the list of theoretical tools. In both cases (a) and (b) I am following the usage of Bohm and Pines.

of Quantum Mechanics. As it turned out, implementing the strategy was a different matter. The authors tell us (617a):

The problem of finding the proper form of S to realize our programme was solved by a systematic study of the equations of motion. We do not have the space to go into the details of this study here but confine ourselves to giving the correct transformation below.[46]

Evidently, in this instance no ready-to-hand tool was available that would do the job.

7. (617b) To continue the narrative of example 6: Given the generator S, H^5 can be transformed into H_{new} by the rule used in example 6: $H_{new} = UH^5U^{-1}$, where $U = \exp(iS/\hbar)$, and $U^{-1} = \exp(-iS/\hbar)$. A certain amount of algebraic drudgery would then yield an expression for H_{new}. Bohm and Pines obtain considerable simplifications, however, by using a standard mathematical tool, another theorem of the algebra of operators:[47]

Given (self-adjoint) operators A, A', and S, if $A' = UAU^{-1}$, where $U = \exp$ (iS), then

$$A' = A - i[A, S] - \tfrac{1}{2}[[A, S], S] + (i/3!)[[[A, S], S]S] + ...$$

Here $[A, S]$ is the *commutator* of A and S, i.e. $[A, S] = AS - SA$.[48]

8. (623b) The final example comes from Appendix I to BP III, in which Bohm and Pines develop a quantum-mechanical version of the approach used in BP II, based on fluctuations of the charge density in the plasma. Here the recourse to the toolbox of theory is explicit: 'We use the electron field second-quantization formalism ... Following the usual treatments,* we describe electrons by the field quantities $\psi(x)$ which satisfy the anti-commutation relations.' At the point marked with an asterisk the authors again cite G. Wentzel, *Quantum Theory of Wave Fields* (1949).

These tools are not all of one kind. They include mathematical identities (examples 6 and 7), existing results within electromagnetic theory, quantum

---

[46] The reader is given no further information about this 'systematic study', and is left to conjecture that it involved a sophisticated form of trial and error.

[47] While the theorem itself is standard, and its proof is straightforward, its use in this context required considerable ingenuity.

[48] Since both the product and the difference of two operators are also operators (hence the expression 'the algebra of operators'), all the expressions on the series on the right-hand side of the equation are well formed.

field theory, and orthodox quantum mechanics (example 2, examples 3 and 8, and example 5, respectively), well-established perturbation techniques of approximation (example 4), and standard mathematical expressions for physical quantities (example 1). As my footnotes and the authors' own citations show, all these tools are to be found in well-known textbooks.

If, as I suggested in the Preamble to this essay, the example of the Bohm–Pines quartet can be generalized, a preliminary account of theoretical practice emerges.[49] To apply, say, quantum mechanics to a particular physical situation—in Kuhnian terms, to work within the paradigm that quantum mechanics provides—a physicist must have a working knowledge of how and when to use the following elements: the simple models that the theory deals with; the mathematical representations of their behaviour that the theory prescribes; the mathematical theory within which these representations are embedded (in this case the theory of Hilbert spaces, which includes as a sub-theory the algebra of operators); and the perturbation techniques and methods of approximation associated with the theory. This congeries of elements I call a *theoretical manifold*. I do not claim that my list of elements is exhaustive. An additional requirement, fifty years after the Bohm–Pines papers were published, is a knowledge of how to use computer methods to assist the work of—or to replace—one or more of the above.

Phrases like 'that the theory deals with' and 'that the theory prescribes', which qualify some of the items in this catalogue, might suggest that underpinning every theoretical manifold there is a theory separable from the manifold it supports. From the perspective of the practitioner of physics, however, that assumption would be a mistake. As Gilbert Ryle said in another context ([1949] 1963: 18):

The same mistake would be made by a child witnessing the march-past of a division, who, having had pointed out to him such and such battalions, batteries, squadrons, etc., asked when the division was going to appear. He would be supposing that a division was a counterpart to the units already seen, partly similar to them and partly unlike them. He would be shown his mistake by being told that in watching the battalions, batteries and squadrons marching past he had been watching the division marching past. The march-past was not a parade of battalions, batteries, squadrons *and* a division; it was a parade of the battalions, batteries and squadrons *of* a division.

---

[49] I augment this account in Section 2.3.1 of the essay.

Analogously, all the items in the theoretical manifold may be thought of as elements *of* the theory; to use a theory is to use one or more of the elements of the theoretical manifold. On this reading, the terms 'theory' and 'theoretical manifold' are coextensive. Call this the *broad* reading of 'theory'.

Customary usage, however, suggests that some elements of the theoretical manifold are more central than others. In the list given, the phrase 'associated with the theory' carries the connotation that the items it qualifies, the theory's 'perturbation techniques and methods of approximation', occupy positions peripheral to the manifold's central core. Central to the manifold, on this account, would be the elements to which the perturbations and approximations are applied—namely, the simple models of the theory and the mathematical representations of their behaviour. We may reserve the term 'theory' for this central cluster of elements, while pointing out that without the other elements of the manifold very little could be achieved in the way of theoretical practice. Call this the *narrow* reading of 'theory'.

I have so far restricted my use of the term 'theory' to foundational theories, theories like classical mechanics, classical electromagnetic theory, or quantum mechanics that have a wide range of applicability. But there is also a more local, but perfectly respectable, use of the term, as when we talk of 'the Bohm–Pines theory of electronic behaviour in metals'. Indeed, Ziman gives this local use primacy. He writes: 'A theory is an analysis of the properties of a hypothetical model.'[50] A *local theory* will employ the theoretical manifold of an existing foundational theory, but will confine its application to a single model, here the jellium model. But, in addition to the standard methods of approximation and perturbation theory that come with the foundational manifold, the local theory may also introduce approximative techniques of its own. Again, the Bohm–Pines theory is a case in point, as we shall see.

2.2.4. Modes of Description

The title of the BP quartet promises a 'Collective Description of Electron Interactions'. In fact, two modes of description, physical and mathematical, are presented. The physical descriptions are in English, augmented by the

---

[50] The quotation comes from Ziman's preface to his *Principles of the Theory of Solids* (1964, v).

vocabulary of physics. Couched in terms of 'electrons', 'fields', and so on, they are, strictly speaking, descriptions of the jellium model, but in the authors' discourse the distinction between model and physical system is rarely observed. In Pines' words, the model chosen is to be 'simple, yet realistic', and the physical descriptions are clearly intended to be construed realistically.

The mathematical descriptions of the model are provided by its Hamiltonian, as we have seen. A Hamiltonian is standardly expressed as a sum of individual terms, each of them either a standard issue Hamiltonian from the toolkit of theory, or obtainable from one by minor modifications. Recall the first example of the use of theoretical tools in BP III, in which Bohm and Pines wrote down a Hamiltonian for their model, 'an aggregate of electrons embedded in a background of uniform positive charge'. The Hamiltonian contained three terms; the first denoted the kinetic energy of the electrons, the second the energy due to the Coulomb interactions between them, and the third their self-energy. The last two were both slightly modified to take into account the uniform background of positive charge. The requirement that the model be 'simple, yet realistic' can be read with such procedures in mind. The simplicity of the model may be judged by the ease with which the Hamiltonian can be constructed from the standard expressions for energy that textbooks provide, its realism by the degree to which the energies associated with the model match those of the system that it represents.

Once the Hamiltonian has been rewritten in collective coordinates, a comparable procedure takes place in the reverse direction. The transformed Hamiltonian is manipulated and approximations are applied until, like the original Hamiltonian, it appears as the sum of recognizable elements, each of them capable of physical interpretation. Thus, as we saw in the synopsis of BP I, the collective approximation allows the transformed Hamiltonian $H_{(1)}$ to be expressed as the sum of three parts. Bohm and Pines write (BP I: 626b):

$$H_{(1)} = H^{(1)}{}_{\text{part}} + H_{\text{osc}} + H_{\text{part int}}$$

and interpret $H^{(1)}{}_{\text{part}}$ as 'the kinetic energy in these new coordinates', H_{osc} as 'a sum of harmonic oscillator terms with frequencies given by the dispersion relation for organized oscillations', and $H_{\text{part int}}$ as a term that 'corresponds to a screened force between particles' (ibid.).

Likewise, in BP III, the authors write:

Let us ... neglect U, a procedure which we have called the random phase approximation ... With this approximation we see that the third and fourth terms in our [transformed] Hamiltonian reduce to

$$H_{osc} = -\tfrac{1}{2}\Sigma_{k<k(c)}(p_k p_{-k} + \omega_p^2 q_k q_{-k})$$

the Hamiltonian appropriate to a set of harmonic oscillators, representing collective fields, with a frequency ω_p.   (BP III: 612b)

In similar vein the other three terms in the Hamiltonian H^4 are interpreted as representing 'the kinetic energy of the electrons', 'a simple interaction between the electrons and the collective field', and 'the short range part of the Coulomb interactions between the electrons' (ibid.); see the synopsis of BP III.

The two modes of description, the physical and the mathematical, while in many ways autonomous, are each responsive to the demands made by the other. They are not totally intertranslatable. Not every describable model can be represented mathematically; not every mathematical description can be given a physical interpretation. Models may need to be simplified; Hamiltonians may need to be mathematically massaged. In both cases the aim is the same: to bring the description closer to the canonical examples supplied by physics textbooks and vice versa. These examples license movement from one mode of description to the other: the rendering of physical descriptions in mathematical terms, and the interpretation of mathematical expressions in physical terms.[51]

This reliance on a comparatively small repertoire of examples imposes a severe constraint on theoretical practice. Or so one might think. But, working within constraints may bring its own benefits. Was sonata form an impediment to Mozart? Or the form of the sonnet a hindrance to Petrarch? In the case at hand, the constraint is positively beneficial—in two ways. In the first place, it solves the *Meno* problem: 'How will you enquire, Socrates, into that which you do not know? ... And if you find what you want, how will

---

[51] The role of these canonical models as 'bridge principles' has been stressed by Nancy Cartwright (1983: Essay 7). She emphasizes the need for them in what she calls 'theory entry', or, in my vocabulary, the move to the mathematical mode of description. As the BP quartet shows, they are also needed for moves in the opposite direction, from the mathematical to the physical mode. For Cartwright (1983: 139), the prime virtue of these examples is explanatory.

you ever know that this is the thing which you did not know?'[52] Bohm and Pines know very well what they are looking for. They seek a Hamiltonian of a recognizable kind, and the standard examples provide aids to recognition. Secondly, the canonical status of these examples allows them to be bearers of meaning. For the habitual reader of the *Physical Review* the interpretations of these mathematical expressions do not have to be re-established *ab initio* on each occasion of their use. Furthermore, the expressions remain meaningful even when they are modified to fit specific circumstances, as in BP III, for example, when the terms in H^1 for the electrons' self-energy and the energy due to Coulomb interactions are modified to take into account the uniform background of positive charge.

Each of the physical descriptions I have so far considered is isomorphic to a corresponding mathematical description. It contains as many clauses as there are terms in the Hamiltonian, each attributing a particular type of energy to the model. But, Bohm and Pines also furnish physical descriptions of another kind. Within the quartet they use the adjective 'physical', together with its cognate adverb 'physically', a dozen times, four times in BP II, seven times in BP III, and once in P IV. On only three occasions is it used in a way that tallies with the account I have given of the interplay between mathematical and physical descriptions of the model. In BP III (619a), for instance, we read: 'The physical consequences of our canonical transformation follow from the lowest-order Hamiltonian $H_{\text{new}}^{(0)}$ ... and the associated set of subsidiary conditions on our system wave function.' More often—in fact on seven of the eleven occasions when the adjective is used in BP II and BP III—it appears in the phrase 'physical picture'. The picture in question is described in the concluding section of BP II (350b–1a).

In conclusion, we give a brief summary of our results in terms of a physical picture of the behavior of the electron gas. As we have seen, the density fluctuations can be split into two approximately independent components, associated, respectively, with the collective and individual particle aspects of the assembly. The collective component, which is present only for wavelengths $> \lambda_D$, represents organized oscillations brought about by the long-range part of the Coulomb interaction. When such an oscillation is excited, each individual particle suffers a small perturbation of its velocity and position arising from the combined potential of all the other particles. The contribution to the density fluctuations resulting from these perturbations

---

[52] Plato, *Meno*, 80d.

is in phase with the potential producing it, so that in an oscillation we find a small organized wavelength perturbation superposed on the much larger random thermal motion of the particle. The cumulative potential of all the particles may, however, be considerable because the long range of the force permits a very large number of particles to contribute to the potential at a given point.

The individual particles component of the density fluctuation is associated with the random thermal motion of the particles and shows no collective behaviour. It represents the individual particles surrounded by co-moving clouds which screen their field within a distance $\sim \lambda_D$. Thus, it describes an assembly of effectively free particles interacting only through the short-range part of the Coulomb force. The screening of the field is actually brought about by the Coulomb repulsion which leads to a deficiency of electrons in the immediate neighbourhood of the particle. This same process also leads to a large reduction in the random fluctuations of the density in the electron gas for wavelengths larger than λ_D.

A third paragraph examines further the interactions of an individual electron with the electron gas as a whole. What is offered here, the authors tell us, is 'a brief summary of [their] results in terms of a physical picture of the behavior of the electron gas'. The 'physical picture' is thus a supplement to the theoretical investigations that have been the main task of the paper. It provides, in a vocabulary markedly different from the one used in descriptions obtained from the system's Hamiltonian, a summary of results already achieved. Organized oscillations are *brought about* by the long-range part of the Coulomb interaction. Each individual particle *suffers* small perturbations *arising from* the combined potential of the other particles. Co-moving clouds *screen* the field of each particle. This screening is *brought about* by the Coulomb repulsion, which *leads to* a deficiency of electrons in the neighbourhood of the particle. The same *process* also *leads to* a large reduction in the random fluctuations of the density in the gas at large wavelengths.

Though it appears late in the second paragraph, the key word here is 'process'. The behaviour of the electron gas is described in terms of causal processes, whereby one thing *brings about*, or *leads to* another. As I have said, the description is presented as a supplement to the theoretical investigations pursued in the quartet. Although its themes are anticipated in the central, theoretical sections of BP II, only once (348a) does the phrase 'physical picture' occur in those sections. Elsewhere in BP II it appears only in the Abstract, in section I: *Introduction*, and section VII: *Conclusion*; similarly, in

BP III, it appears twice in the first section, once in the last, but nowhere else. This points to the fact that the picture itself is relatively independent of the theoretical approaches taken. Though drawn from the classical account of the electron gas given in BP II, it nevertheless holds good alongside the quantum mechanical account in BP III. Descriptions of this kind—'narrative descriptions', as I will call them—are used throughout physics. Although their closest affiliation is with the theoretical manifolds of classical physics, within theoretical practice they are effectively independent of 'high theory', and, as in the present case, can coexist alongside theoretical manifolds of many different persuasions. And, for obvious reasons, they are a large part of the lingua franca of experimental practice.[53]

2.2.5. The Use of Approximations

In just five pages of BP I (sections II C and II D, and section III B), the verb 'to neglect', variously conjugated, appears no fewer than twelve times. In these sections, Bohm and Pines are examining the results of moving from individual to collective coordinates, first in the classical case and then in the quantum case. In sections II A and III A, respectively, the transformations that will effect these moves have been specified with mathematical exactness. After they are applied to the original Hamiltonian H_0, however, approximative strategies are brought into play, so that terms that make only small contributions to the Hamiltonian disappear. In this way, the otherwise unwieldy expression for the transformed Hamiltonian is presented in a form that allows each of its components to be given a physical interpretation.

Similar approximations are used throughout the quartet. Each involves the assumption that some aspect of the physical situation guarantees that corresponding terms in its mathematical description can be neglected without grossly affecting the results of the analysis. These assumptions are brought together and discussed in section II B of BP I (628a–b) under the title 'The Collective Approximation'. There are four of them; I will comment briefly on each in turn.

[53] The importance of narrative descriptions has been emphasized by several authors. Stephan Hartmann (1999) has drawn attention to the contribution they make to hadron physics. He calls them simply 'stories'. The narrative mode also undergirds a different account of physical science. Peter Machamer, Lindley Darden, and Carl Craven (2000) have shown that our best account of neuroscience is in terms of *mechanisms*, and the descriptions they provide of mechanisms are narrative descriptions.

1. Electron–ion and electron–electron collisions are ignored. I drew attention to this simplification in section 2.2.2. It is of the same kind as the simplification (some philosophers of physics call it 'abstraction') whereby the effect of air resistance on the motion of a pendulum is neglected (see n. 31). In each case, the effect of the neglected element is to produce a small damping of the oscillations of the system. The neglect of electron–ion collisions is not one of the approximations used in BP I to simplify the transformed Hamiltonian; rather, it is implicit in the authors' choice of the jellium model, in which the effect of the plasma ions is represented by a background of uniformly distributed positive charge.

2. The organized oscillations in the electron gas are assumed to be small. The 'customary linear approximation, appropriate for small oscillations' (BP I: 628b) is then used, allowing quadratic field terms (i.e. products of field terms) in the equations of motion of the system to be neglected.

3. In BP I, the velocity v of the electrons is assumed to be small compared with c, the velocity of light, so that terms involving v^2/c^2 are negligible. (This approximation appears repeatedly; it accounts for eight of the twelve occurrences of 'neglect' that I mentioned earlier.) In BP III, as we have seen, the assumption is made that there is a maximum wave number k_c (equivalently, a lowest wavelength λ_c) for organized oscillations. These assumptions are presented as two versions of a single assumption, that $(\mathbf{k.v})/\omega$ is small, appropriate for transverse and longitudinal oscillations, respectively.

These first three approximations are straightforward, the last, the 'random phase approximation', less so.

4. Bohm and Pines write (BP I: 628b):

We distinguish between two kinds of response of the electrons to a wave. One of these is in phase with the wave, so that the phase difference between the particle response and the wave producing it is independent of the position of the particle. This is the response which contributes to the organized behaviour of the system. The other response has a phase difference with the wave producing it which depends on the position of the particle. Because of the general random location of the particles, this second response tends to average out to zero when we consider a large number of electrons, and we shall neglect the contributions arising from this. This procedure we call the 'random phase approximation'.

The claim here is not that out-of-phase responses are individually negligible—they may well be of the same order of magnitude as the in-phase responses. The assumption made is that, taken collectively, they cancel each other out. Bohm and Pines justify the r.p.a. mathematically in the classical treatment of longitudinal oscillations given in BP II (349–50), and in 'a more qualitative and physical fashion' in the quantum mechanical treatment given in BP III (621).

The second of these approximative strategies and the first part of the third are standard fare, and I will say more about them in section 2.2.7. I call them 'strategies' because their application takes different forms in different theoretical contexts. None of the four is specific to the particular theoretical manifolds that the authors use. That is why the authors need to provide an extended discussion of them. As they point out (BP I: 628b), this four-part collective approximation differs from an orthodox perturbation theory in that the latter would not allow for the fact that small changes in the field arising from each particle may add up to a large change in the net field—which, on the authors' account, is precisely what generates plasma waves in metals.

2.2.6. Experiment and the Bohm–Pines Theory

Within the Bohm–Pines quartet, little attention is paid to experimental findings, none at all to experimental practice. The first paper deals with the possibility of transverse oscillations due to electromagnetic interactions between moving electrons, the other three with the possibility of longitudinal oscillations due to electrostatic (Coulomb) interactions. The role of BP I is 'to illustrate clearly the techniques and approximations involved in [the authors'] methods' (BP I: 627a). Virtually no mention is made of observed phenomena or experimental results. Bohm and Pines tell us: 'The electromagnetic interactions [which give rise to transverse oscillations] are weaker than the corresponding Coulomb interactions by a factor of v^2/c^2 [where c is the velocity of light] and, consequently, are not usually of great physical interest' (ibid.).

Within the papers that follow, discussion of observable phenomena and experimental results is confined to BP II and P IV. BP III is entirely theoretical; the only mention of a possibly observable phenomenon comes in an aside (BP III: 610a–b) where the authors cite the work of two other theorists (Kronig and Kramer) who 'treated the effects of electron–electron interaction on the stopping power of a metal for fast charged particles'.

In BP II and P IV, a related phenomenon, the loss of energy suffered by a high-energy electron when it passes through a metal film, is adduced as evidence to support the authors' own analysis. In a paragraph of the Introduction to BP II, the authors predict the excitation of collective oscillations in a metal by high-energy particles passing through it; the paragraph ends with a laconic reference to experimental confirmation: 'Experiments by Ruthemann and Lang, on the bombardment of thin metallic films by fast electrons tend to verify our theoretical predictions concerning this type of oscillation' (BP II: 339a).

This remark is amplified in section IV of the paper, where the link between theory and experiment is provided by a 'correspondence principle argument' (see sections 2.1.3 and 2.2.3 of this essay). In section IV of P IV, Pines uses quantum mechanics to provide an explanation of this phenomenon, which is summarized in the penultimate paragraph of section 2.1.5 of this essay. Analysis of the interaction of the high-energy electrons with the collective field shows that only at one particular frequency ω, very close to the plasma frequency ω_P, will the oscillations of the field become self-sustaining (P IV: 634a). Pines points out (P IV: 634b) that for aluminium and beryllium films the predicted values of energy lost (in multiples of $\hbar\omega$) by the electrons agree very well with those obtained experimentally by Ruthemann and by Lang. Furthermore, the calculated mean free path (the average distance travelled by the high-energy particle between excitations) matches an empirical value based on Lang's data for aluminium films.

When gold, copper, or nickel films are used, however, the spectrum of energy loss is not discrete. Pines attributes this failure to the fact that 'the valence electrons in these metals are not sufficiently free to take part in undamped collective motion' (ibid.)—that is, that the jellium model, in which there is no interaction between valence electrons and the individual ions of the metal, is inadequate for these elements. He notes, 'Experiments have not yet been performed on the alkali metals, where we should expect to find collective oscillation and the appearance of discrete energy losses' (ibid.).[54]

There are four other places in P IV where empirical evidence is referred to. Three instances occur in section III, where Pines is discussing the surprising fact that, if the independent electron model of electron behaviour is 'corrected' to allow for the so-called 'exchange energy' between electrons, the result is to worsen the agreement with experiment rather than to improve it (see section 2.1.5 of this essay). One instance involves the magnetic properties of

[54] Alkali metals lithium, sodium, potassium, etc. appear in Group Ia of the Periodic Table. They have just one valence electron per atom.

the electron gas, another the way its specific heat varies with temperature. In neither case does Pines cite the experiments that yielded the relevant empirical results. Nor does he in the third instance (P IV: 631a). In this instance Pines describes an *ad hoc* theoretical move made by P. T. Landsberg. In order to account for an observed feature of the x-ray emission spectrum for sodium, Landsberg had found it necessary to introduce a screened Coulomb interaction between electrons. Since a screening effect of this kind is predicted by the BP theory, one might expect Pines to present Landsberg's manoeuvre as an indirect confirmation of that theory. Instead, Pines points to a result by E. P. Wohlfahrt. Landsberg had proposed that the screened Coulomb interaction could be mathematically modelled by writing $(e^2/r_{ij})\exp[-(r_{ij}/\lambda)]$ for the electron–electron interaction potential in place of the standard expression e^2/r_{ij}, and that the 'screening radius' λ was of the order of 10^{-8} cm. Wohlfahrt showed that if both of these proposals were accepted, then the unfortunate effect of 'correcting' the independent electron model to allow for the exchange energy would be greatly reduced, along with the error in its predictions concerning specific heat. The comparison implicit in Pines' discussion of Landsberg's and Wohlfahrt's work is this: on the independent electron account of a plasma there are results that can be purchased only by making *ad hoc* assumptions concerning the electron–electron interaction potential within the plasma. On the BP approach, in contrast, they come for free.

Here, as throughout the first three sections of P IV, Pines is more concerned to compare the BP theory with other theoretical approaches than with experimental results. In addition to the work of the theorists I have already mentioned, Pines points to two separate treatments of an electron gas by Eugene Wigner (1934; 1938), and shows how their results match those obtained using the collective approach. This form of validation, whereby a new theoretical approach gains credence by reproducing the results of an earlier one, has received little attention from philosophers of science, though instances of it are not far to seek. For example, when Einstein first proposed his general theory of relativity, one of the earliest tasks he set himself was to show how the theory could recapture, to a first approximation, the classical principle of conservation of energy as it applied to an orbiting planet.[55] In the present case, a general sense that, for Bohm and Pines, empirical justification

---

[55] It appears in Einstein 1915, the paper in which he explained the anomalous advance of the perihelion of Mercury. This paper is the third in a quartet of papers, all published in Nov. 1915, in which Einstein introduced the general theory.

was something of an afterthought is reinforced by an acknowledgement they offer at the end of BP II (351a): 'The authors wish to thank Dr Conyers Herring for informing us of the experiments of Ruthemann and Lang.'

Be that as it may, it is certainly true that, in the BP quartet, theory and experiment exist as almost independent disciplines, one of them barely glanced at. We read that both Ruthemann and Lang conducted experiments in which high-energy electrons were directed at thin metallic films and their energy losses in passing through the films were measured. We are told (P IV: 634b) that in Lang's experiments the initial energy of the electrons was 7.6 ev, and that the energy losses were multiples of 14.7 ev in aluminium films and 19.0 ev in beryllium films. (The calculated values are 15.9 ev and 18.8 ev, respectively.) But, we are told neither how these beams of electrons were prepared, nor how the electrons' initial energies and losses in energy were established. It seems that the practices involved in preparation and measurement belong exclusively to the guild of experimenters, and are not the province of the worshipful company of theoreticians. But this cannot be the whole story. While the two sets of practices, experimental and theoretical, may be different, they cannot be wholly disjoint, for two reasons. The first is that experimental practices include theoretical practices—a truism which has the virtue of being true. The energy of an electron may be measurable, but it is not observable, and every measuring instrument more sophisticated than a metre rule is a material realization of a theory. When a theoretical result is compared with an experimental result, the practices involved in the theoretical side of things may or may not belong within the same theoretical manifold as those on the experimental side. Although two different theoretical manifolds are used in BP II and P IV, in both cases the conclusions reached are compared with the results of the same experiments; thus, in at least one of these cases the theoretical manifold used by Bohm and Pines differs from that used by Lang. None the less—and here I come to the second point—if empirical confirmation is to be possible, in both cases the set of practices on the theoretical side must mesh with the set of practices on the empirical side. Theoretician and experimenter must agree that a theoretical energy loss of 18.8 ev can be meaningfully compared with an experimental energy loss of 19.0 ev. In both the literal and the Kuhnian sense, the two sets of practices cannot be incommensurable. In Galison's phrase (1997 *passim*), the two parties need to establish a *trading zone*. How this might be done is beyond the scope of this book.

The fourth and last instance of an occasion when empirical evidence is at issue is an instance of a slightly peculiar kind. It occurs very early on in P IV (626–7) where, as in the third instance, Pines is comparing the collective account of electron behaviour with the account given by the independent electron model. On these occasions Pines is in the position of one who needs to explain the curious incident of the dog in the night time. As readers of Conan Doyle will recall, what made the incident curious was that the dog did nothing in the night time. In like manner, Pines must explain the fact that the collective behaviour that he and Bohm ascribe to metallic plasmas has very little effect on the plasma's properties and behaviour, so little that the independent electron model, in which 'the motion of a given electron is assumed to be independent of all the other electrons', enjoys widespread success. He makes this explanation his first order of business; a qualitative explanation is given in the paper's introduction, and a quantitative account follows in its central sections (see section 2.1.5 above). In certain circumstances, the absence of a phenomenon may stand as much in need of explanation as would the phenomenon itself.

2.2.7. Deduction in the Bohm–Pines Quartet

I introduced Part 2.2 of this essay by quoting John Ziman's delineation of the intellectual strategy of a typical paper in theoretical physics: 'A model is set up, its theoretical properties are deduced and experimental phenomena are thereby explained.' Taking my cue from the first clause of that account, I began my observations on the BP quartet by describing the role played by models and modelling. Turning my attention to the second clause, I will end them by examining the notion of *deduction*, as it appears in the third of these papers.

A curious feature of Chapter 3 of Ernest Nagel's *The Structure of Science* (1961) is that, although it bears the title 'The Deductive Pattern of Explanation', there is no mention in it of the steps by which conclusions follow from premises. In that chapter, the actual process of deduction is taken for granted, witness this paragraph (ibid. 32).

[A] deductive scientific explanation, whose explanans is the occurrence of some event or the possession of some property by a given object, must satisfy two logical conditions. The premises must contain at least one universal, whose inclusion in the premises is essential for the deduction of the explanandum. And the premises must also contain a suitable number of initial conditions.

This is a beautiful example of what, borrowing the phrase from Nancy Cartwright (1999: 247), I will call the 'vending machine' view of theorizing. Originally, Cartwright used the phrase in criticizing a particular account of theory entry, the process by which, in Ziman's words, 'a model [of a physical system] is set up': 'The theory is a vending machine: you feed it input in certain prescribed forms for the desired output; it gurgitates for a while; then it drops out the sought-for representation; plonk, on the tray, fully formed, as Athena from the brain of Zeus' (Cartwright 1999: 247).

The feature of the machine that Cartwright challenges is its mode of input, which must accord with 'certain prescribed forms' (ibid.).[56] I will use the same metaphor to characterize the next stage of Ziman's narrative, where 'theoretical properties of the model are deduced and experimental phenomena are thereby explained'. In this usage, the salient feature of the machine will be the process of *gurgitation*. In chapter 2 of this book, Nagel talked (albeit unknowingly) about that phase of the machine's working (1961: 21): 'A type of explanation commonly encountered in the natural sciences ... has the formal structure of a deductive argument, in which the explanandum is *a logically necessary consequence* of the explanatory premises' (my emphasis). Here, 'logical necessity' includes 'mathematical necessity', as Nagel points out a few lines later. This kind of deduction we may call *strict deduction*; it is the kind of deduction that Nagel subsequently takes for granted.[57]

Pace Nagel, adherence to strict deduction is not 'commonly encountered in the natural sciences'. An insistence on strictness would mean, for example, that the explanation of the behaviour of a system consisting of more than two mutually attracting bodies would be beyond the scope of classical mechanics. Starting with Newton, physicists have become adept at finding ways around this problem, and Bohm and Pines are no exception. To illustrate how they go about it, I will examine the extended argument in BP III that transforms the Hamilton for the jellium model from H^1 to H_{new}, and in so doing makes manifest the collective properties of the electron gas. I will first give a precis of the argument, and then list the individual moves within it. The order

---

[56] She continues, 'Producing a model of a new phenomenon such as superconductivity is an incredibly difficult and creative activity. It is how Nobel prizes are born' (ibid.). The essay from which this quotation is taken deals at length with the Bardeen, Cooper, and Schrieffer theory of superconductivity for which they were indeed awarded the Nobel prize. I give a brief account of their achievement in section 2.3.1 of this essay.

[57] See, for instance, the extended footnote on p. 353 of *The Structure of Science*.

in which they are listed, however, is determined, not by the order of their occurrence in BP III, but by the kinds of justification offered for them. First in the list are moves justifiable by strict deduction; next come moves that have obvious, but not strictly deductive, justifications; at the end of the list are moves that are highly pragmatic, and are peculiar to the investigation under way. By now virtually all these moves will be familiar to the reader. My aim in setting them as I do is to draw attention to the gulf between the concept of 'deduction', as the word is used by Ziman, a working physicist, and as it is used by Nagel, a mid-twentieth-century philosopher of science.

First, the precis, in six steps:

Step (1) Bohm and Pines present a standard Hamiltonian H^1 for a cloud of electrons against a background of uniform positive charge.

Step (1 → 2) H^1 is then claimed to be equivalent to H^2, given a family of subsidiary conditions Ω_k on the state of the system. The claim is justified by displaying a unitary operator S such that $H^2 = SH^1S^{-1}$. Two central terms in H^2 are expressed as Fourier series, indexed by k (as was the term for the energy due to Coulomb interactions in H^1), and each of the subsidiary conditions corresponds to a component k of the Fourier decompositions.

Step (1 → 3) The procedure of Step (1 → 2) is now repeated, using a modified operator S, so that the two Fourier series in the terms of the resulting Hamilton, H^3, are truncated; neither of them has an index higher than k_C. Correspondingly, there are just k_C subsidiary conditions Ω_k.

Step (3 → 4) Algebraic manipulation of H^3 yields the Hamilton H^4, which contains five terms.

Step (4 → 5) This step contains just two moves. One of the terms of H^4, $U^\dagger$, is shown to be negligible, and is therefore deleted;[58] another, H_{osc}, is rewritten to produce H_{field}. The combined result is H^5.

Step (5 → new) Bohm and Pines then apply a 'canonical transformation' to H^5 (and to the operators Ω_k), and obtain the final Hamilton, H_{new}.

---

[58] Recall that within BP III, Bohm and Pines are not consistent in their use of the letters 'U' and 'S'. In Step 5 →*new* the letters refer to a unitary operator and its generator, respectively. The authors have, however, previously used 'U' to refer to a component of the Hamiltonian H^4, and 'S' to refer to a unitary operator in Steps 1 → 2 and 1 → 3. In this section I resolve one of these ambiguities by writing '$U^\dagger$' for the component of H^4.

Now for the list of moves categorized by justification; the page numbers cited are all from BP III.

Trivial moves

Step $(3 \rightarrow 4)$ involves just one move, an elementary example of strict deduction, in which H^4 is obtained by rearranging the components of the terms of H^3.

Strictly mathematical moves

1. In Step (1), the authors 'have used the fact that the Coulomb interaction between the ith and j^{th} electrons may be expanded as a Fourier series in a box of unit volume' (610a). Though their subject matter is the physical world, the 'fact' they use is a mathematical fact, belonging to the branch of mathematics known as analysis.

2. Likewise, in Step $(4 \rightarrow 5)$, when the field operators q_k and p_k are expressed in terms of creation and annihilation operators a_k and $a_k{}^*$ (616a), although this move has a physical interpretation, it is a perfectly permissible formal move, made 'to simplify the commutator calculus'.

3. As we have seen, Steps $(1 \rightarrow 2)$, $(1 \rightarrow 3)$, and $(5 \rightarrow new)$ each involves a unitary transformation of a Hamilton H^a into another, H^b, In essence, a transformation of this kind is equivalent to a move from one system of coordinates to another in the (abstract) Hilbert space on which these operators are defined. Formally, these transformations employ a mathematical identity in the operator algebra of that Hilbert space of the kind $H^b = UH^aU^{-1}$ (see example 6 in section 2.2.3).

In Step $(5 \rightarrow new)$, after a move of type 3, Bohm and Pines use another theorem of the operator algebra to rewrite the right hand side of that identity:

$$H^b = UH^aU^{-1} = H^a - i[H^a, S] - \tfrac{1}{2}[[H^a, S], S]$$
$$+ (i/3!)[[[H^a, S], S], S] + \ldots$$

Here S is the *generator* of U: $U = \exp(iS/h)$. The term '$[H^a, S]$' is a commutator of the first order, '$[[H^a, S], S]$' a commutator of the second order, and so on.

Standard approximations

Physics, in particular the physics of many-body systems, is not an exact science. Approximations (in the literal sense: numerical results 'very close' to the 'real' values) are not only tolerated but inevitable. If a formula for a physical quantity Q contains a term t whose value is small in comparison with the value of Q, then a term t^2 of the second order will often be neglected. In quantum mechanics, a similar procedure applies to the operator representing Q. Two of the four types of approximation discussed in BP I (and in section 2.2.5 above) conform to this pattern: neglected are quadratic field terms and the term v^2/c^2, when v is small compared with c. Step ($5 \rightarrow$ *new*) in BP III provides an example of the former kind (618a): 'in obtaining [equation (54)] we have neglected a number of terms which are quadratic in the field variables and are multiplied by a phase factor with a non-vanishing argument'. Another example from the same step involves the mathematical move we have just encountered, whereby a unitary transformation is unpacked as a series of commutators involving the generator S of the relevant unitary operator:

$$H^b = UH^aU^{-1} = H^a - i[H^a, S] - \tfrac{1}{2}[[H^a, S], S]$$
$$+ (i/3!)[[[H^a, S], S], S] + \ldots$$

Paradoxical though it may seem, the use of this identity leads to considerable simplifications. In preparation for that move, Bohm and Pines have defined an *expansion parameter*, α, which is a measure of the strength of 'the coupling between the field and electrons' (615b). They expected α to be small; in fact they described it as 'the measure of the smallness' of a term in the system's Hamiltonian (ibid.). If we replace H^a in the identity above by H_I (one of the terms of H^4) it transpires that 'the effects of the field-particle interaction (up to order α) are contained in the first correction term, $(i/\hbar)[H_I, S]$. The higher-order commutators will be shown to be effects of order $\alpha^2, \ldots$ *and may hence be neglected*' (618a; my emphasis; $\hbar$ is Planck's constant).[59] To anticipate: in the vanishing case, when $[H^a, S] = 0$, then $H^b = H^a$, and H^a is unaffected by the transformation.

---

[59] A similar strategy was adopted in BP I. At the corresponding point in their quantum mechanical treatment of *transverse* oscillations, Bohm and Pines tell us that 'higher order commutators can be neglected if we restrict our attention to the lowest order terms in v/c' (632b). I will come back to the commutator $i/\hbar[H_I, S,]$ later.

Perturbation theory

In quantum mechanics the treatment of small terms is codified by *perturbation theory*. A basic element of quantum theory is the (time-independent) Schrödinger equation, which relates the Hamiltonian to a spectrum of values of energy. The values are determined by the solutions of the equation. A problem arises when a Hamiltonian has the form $H_a + H_b$, and H_b is small compared with H_a, because in that case the Schrödinger equation may not have exact solutions. Perturbation theory then provides a procedure whereby an approximate solution may be found.

Step $(4 \rightarrow 5)$ provides an example. Bohm and Pines were faced with a Hamiltonian (H^4) which contained five terms: $H_{part} + H_I + H_{osc} + H_{s.r.} + U^\dagger$. Since both $U^\dagger$ and H_I were known to be small compared with the other three terms, Bohm and Pines used second-order perturbation theory to determine just how small those effects were. Bohm and Pines concluded that they were 'justified in neglecting completely the term $U^\dagger$' (BP III: 614b), but that, although they were 'justified in neglecting H_I in order to obtain a qualitative and rough quantitative understanding of the behaviour of [their] system ... the effects arising from H_I should be taken into account in a careful quantitative treatment' (ibid. 615a).[60] Note that perturbation theory does not go beyond orthodox quantum theory; it merely provides a recipe whereby selected terms in the Schrödinger equation are discarded; in first-order perturbation theory, terms of second order are discarded, in second-order perturbation theory, terms of third and higher orders degenerate.

Formal analogies

I turn now to analogical reasoning, specifically to the use within BP III of arguments and results from the two papers that came before it. BP I contained both a classical and a quantum mechanical treatment of transverse plasma waves, and BP II a classical treatment of longitudinal plasma waves. BP III gives a quantum mechanical treatment of longitudinal plasma waves, and so marries the second of the theories used in BP I to the phenomena discussed in BP II. It is therefore not surprising that the correspondences drawn between BP III and BP I are different in kind from those drawn between BP III and BP II.

---

[60] Recall that the transformation that took H^5 into H_{new} was designed to distribute the effects of H_I among those terms that remained. See section 2.2.4 and n. 20.

In BP III, Bohm and Pines reserve the words 'analog(ue)' and 'analogous' for the correspondences between that paper and the quantum mechanical treatment provided in BP I, and, on each of the eight occasions when BP I is mentioned, one at least of those words occurs. The analogies the authors refer to are formal analogies, correspondences between mathematical formulae. For instance, 'H_I and H_{field} are ... analogous to the transverse terms encountered in BP I, and we may expect that many of the results obtained there may be directly transposed to the longitudinal case' (617a). Likewise, the operator S, which is instrumental in the canonical transformation of H^5 to H_{new} in Step (5 → *new*), 'may be seen to be just the longitudinal analogue of the "transverse" generating function given in BP I' (ibid.).

Non-standard approximations

As we have seen, in Step (4 → 5), Bohm and Pines use second-order perturbation theory to justify the neglect of the term $U^\dagger$ in H^4. They had previously used another argument to the same end:

$U^\dagger$ ... always depends on the electron coordinates, and since these are distributed over a wide-range of positions, there is a strong tendency for the various terms entering into $U^\dagger$ to cancel out. Let us for the time being neglect $U^\dagger$, a procedure we have called the random-phase approximation, and which we shall presently justify. BP III: 6126

Physical analogies

Though Bohm and Pines had used two distinct theories in BP I, in each case their approach had been the same. The starting point of each was a Hamiltonian for a collection of electrons in a transverse electromagnetic field in which the electrons were represented by the momentum p_I and position x_I of the individual charges. A similar approach was taken in BP III; the only difference was that the longitudinal rather than the transverse components of the electromagnetic field are included in the Hamiltonian. In both these papers, only after a canonical transformation was applied to the Hamiltonian was the system described in terms of collective variables. In contrast, in BP II Bohm and Pines wrote:

Our approach to the equations of motion is aimed at making use of the simplicity of the collective behavior as a *starting point* for a tractable solution. *As a first step*, we study the fluctuations in the particle density, because ... their behavior provides a good measure of the applicability of a collective description. (BP II: 340a; my emphases)

In other words, since the particle density is a collective property, nothing corresponding to the canonical transformations used in BP I is needed in BP II. And, while it is true that in BP III Bohm and Pines develop 'a direct quantum mechanical extension of the methods used in Paper II', this extension is relegated to a brief appendix (623–4), and is used in section V of the paper only to resolve some complications involving the subsidiary conditions on the quantum states of the system. In the theoretical core of BP III, the mathematical aspects of BP II are set to one side. Instead, Bohm and Pines emphasize the 'physical picture' the paper presented:

In the preceding paper [BP II] we developed a detailed physical picture of the electronic behavior [of a dense electron gas]. Although the electron gas was treated classically, we shall see that most of the conclusions reached there are also appropriate (with certain modifications) in the quantum domain. Let us review briefly the physical picture we developed in [BP II], since we shall have to make frequent use of it in this paper.   (609a)

As we have seen, the most striking feature of the picture was that, within it:

[T]he electron gas displayed both collective and individual particle aspects. ... The collective behaviour [i.e., the plasma oscillations] of the electron gas is decisive for phenomena involving distances greater than the Debye length, while for smaller distances the electron gas is best considered as a collection of individual particles which interact weakly by means of a screened Coulomb force.   (609a–b)

I will focus on just two of the correspondences between BP III and BP II that Bohm and Pines rely on. The first is very general: it involves the demarcation marked in the classical case by the Debye length. The second is more specific: it concerns the *dispersion relation*, the relation between the frequency and the wavelength of plasma oscillations.

1. Some prefatory remarks are needed. In BP II (341a–b), Bohm and Pines present a criterion for the applicability of a collective description. The criterion involves an equation (labelled '(9)'), which has been deduced from the principles of electrostatics with the help of the random phase approximation. In this equation, $d^2\rho_k/dt^2$ is equated with the sum of two terms,[61] one representing the random thermal motion of the individual

---

[61] Recall that ρ is the electron density and k is the wave number of the oscillations. It appears when Bohm and Pines represent the electron density ρ by the Fourier decomposition $\rho = \Sigma_k a_k \rho_k$. See section 2.1.3.

particles, the other the collective oscillations of the electron gas. Thus, the 'rough criterion for the applicability of a collective description' is that 'for most particles the collective [second] term in (9) be much greater than the term arising from the random thermal individual particle motions' (BP II: 341a–b). It turns out that, if k is sufficiently small, the effect of the first term can be neglected, and a straightforward derivation shows that the quantitative form of this condition can be written as $k^2 << 4\pi n\lambda_{D^{-2}}$, where λ_D is 'the well known Debye length' (ibid.).[62] Conversely, for high wave number k (and small wavelength λ), collective behaviour (i.e. plasma oscillations) will not be generated. This result is carried over into Step (1 → 3) of BP III:

We found in Paper II that in the classical theory there is a minimum wave wavelength λ_C (which classically is the Debye length), and hence a maximum wave vector k_C, beyond which organized oscillation is not possible. We may anticipate that in the quantum theory a similar (but not identical) limit arises, so that there is a corresponding limit on the extent to which we can introduce collective coordinates to describe the electron gas. ... The number of collective coordinates, n', will then correspond to the number of k values lying between $k = 0$ and $k = k_C$. ... The modification of $[H^1]$ to include only terms involving (p_k, q_k) with $k < k_C$, may be conveniently carried out by applying a unitary transformation similar to [the transformation which took H^1 to H^2], but involving only [position coordinates] q_k for which $k < k_C$. (BP III: 611b–12a)

By this means, H^1 is transformed into H^3, and the information about plasma behaviour obtained by classical means becomes encoded in a quantum-theoretic Hamiltonian. This correspondence, however, is not quantitative. As the quotation above tells us, Bohm and Pines anticipate that, in the quantum case, the limit k_C will be 'similar, but not identical' to the classical limit, where it is the reciprocal of the Debye length λ_D.

2. The *dispersion relation* of a plasma relates the frequency ω of a plasma oscillation to its wave number k (the reciprocal of its wavelength). The relation is presented on p. 618b, in the penultimate move of Step (5 → *new*). Bohm and Pines show that, because of the form it takes, significant simplifications can be effected.[63] They write: 'This dispersion relation plays a key role in our

---

[62] λ_D is defined by the equation: $\lambda_D = \kappa T/4\pi ne^2$, where κ is Boltzmann's constant and T is the absolute temperature of the electron gas.

[63] To be precise, several terms in the first version of H_{new} disappear if the dispersion relation takes the form suggested (618b). The terms in question are the first two terms of the commutator $i/\hbar[H_1, S,]$ and the transformed version of one of the terms of H_{field} in H^5.

collective description, since it is only for $\omega(k)$ which satisfy it that we can eliminate the unwanted terms in the Hamiltonian ... and the unwanted field terms in the subsidiary condition' (619b).

The fact that a particular choice for the dispersion relation leads to desirable results does not, however, justify that choice, as Bohm and Pines acknowledge. They continue:

The frequency of these collective [longitudinal] oscillations is given by the dispersion relation ... which is the appropriate quantum mechanical generalization of the classical dispersion relation derived in BP II, as well as being the longitudinal analog of the quantum-dispersion relation for organized transverse oscillation, which we obtained in BP I.   (ibid.)

This is a double-barrelled justification, appealing as it does both to a physical and to a formal analogy.

Causal inferences

In this catalogue of argumentative moves, all but one of those under the headings *Non-standard approximations, Physical analogies*, and *Causal inferences* (that is to say, those moves furthest away from the canons of strict deduction) are buttressed by more than one argument.[64] The two arguments under the heading *Causal inference* both involve the term $H_{s.r.}$, which appears in H^4 and H^5 and represents 'the short-range part of the Coulomb interaction between the electrons' (612b). In each case, Bohm and Pines want to establish that the canonical transformation which takes H^5 to H_{new} has a negligible effect on $H_{s.r.}$; or, in mathematical terms, if U is the unitary operator associated with the transformation, then to a good approximation, $UH_{s.r.}U^{-1} = H_{s.r.}$.

1. An example from the strictly mathematical moves shows that the desired conclusion holds, provided that the commutator $[S, H_{s.r.}]$ is negligible (where S is the generator of U). Bohm and Pines present 'a typical first order term arising from $[S, H_{s.r.}]$', and label it '(74)'. They point out that 'the structure of (74) is quite similar to that of $U^{\dagger}$ [in H^4]' (621a). As we have seen,

---

[64] In the non-standard approximation, the neglect of $U^{\dagger}$ was justified by the r.p.a., but further justification was provided, as we have seen, by second-order perturbation theory applied to the single operator $U^{\dagger}$ (614b). Furthermore, justification of the r.p.a. in the classical case appeared in section VI of BP II, and although it was tailor-made for a specific example readers were invited to regard it as a template to be used elsewhere. And, as I pointed out at the end of the discussion of the second physical analogy, the choice of dispersion relation had a twofold justification.

$U^\dagger$ was shown to be negligible by the use of second-order perturbation theory; in treating (74), however, the authors opt for a different strategy (621a–b):

Because of the analytic difficulties involved [in the use of perturbation theory], we prefer to justify our neglect of (74) in a more qualitative and physical fashion.

We see that (74) describes the effect of the collective oscillations on the short-range collisions between the electrons, and, conversely, the effect of the short-range collisions on the collective oscillations. ... The short-range electron-electron collisions arising from $H_{s.r.}$ will act to damp the collective oscillations ... A test for the validity of our approximation in neglecting terms like (74) is that the damping time from the collisions be small compared with the period of a collective oscillation. In this connection we may make the following remarks:

1. Electron-electron collisions are comparatively ineffective in damping the oscillations, since momentum is conserved in such collisions, so that to a first approximation such collisions produce no damping.

2. The exclusion principle will further reduce the cross section for electron-electron collision.

3. If H_1 [which represents 'a simple interaction between the electrons and the collective field'] is neglected, collisions have no effect on the collective oscillations. This means that the major part of the collective energy is unaffected by these short-range collisions, since only that part coming from H_I (which is of order α [the expansion parameter] relative to ω_p) can possibly be influenced. Thus at most 20 percent of the collective energy can be damped in a collision process.

All of these factors combine to reduce the rate of damping, so that we believe this rate is not more that 1 percent per period of an oscillation and probably quite a bit less. ... It is for these reasons that we feel justified in neglecting the effects of our canonical transformation on $H_{s.r.}$.

We have been given two reasons for taking (74) (and hence $[S, H_{s.r.}]$) to be negligible compared with other terms in H^5: first, Bohm and Pines drew attention to the existence of formal similarities between (74) and $U^\dagger$ (a term already known to be negligible); secondly, they presented a causal argument to show that the effects, represented by (74), of interactions between long-range collective oscillations and short-range collisions of electrons would be small.[65]

---

[65] In addition, the authors draw attention to a classical treatment of damping of collective oscillation by electron–electron collisions by Bohm and Gross 1949a and b.

Even before the canonical transformation took place (616b–18b), a condensed version of this argument had been presented (616a):

From [the expressions for the components of H^5] we see that if we neglect H_I, the collective oscillations are not affected at all by $H_{s.r.}$. Thus $H_{s.r.}$ can influence the q_k only indirectly through H_I. But, as we shall see, the *direct* effects of H_I on the collective oscillations are small. Thus, it may be expected that the *indirect* effects of $H_{s.r.}$ on the q_k through H_I are an order of magnitude smaller and may be neglected in our treatment which is aimed at approximating the effect of H_I.

(The variable q_k that appears here has been introduced in H^2. It represents a generic component of the Fourier expansion of **A**, the electromagnetic vector potential in the plasma, and will be supplanted by a collective variable when the canonical transformation takes place.)

I call this argument a 'causal inference' because of the vocabulary employed: '$H_{s.r.}$ can *influence* the q_k only indirectly'; 'the *direct effects* of H_I'; 'the *indirect effects* of $H_{s.r.}$'; 'collective oscillations are not *affected* at all by $H_{s.r.}$.' (emphases mostly mine). The agents portrayed as bringing about effects, directly or indirectly, are components of the Hamilton H^5. But, here we should listen to the stern voice of Pierre Duhem (*PT*, 20): 'These mathematical symbols have no connection of an intrinsic nature with the properties they represent; they bear to the latter only the relation of sign to thing signified.' In what sense, then, can a theoretical term be a causal agent? The short answer is that in this argument Bohm and Pines have moved into a figurative discourse, within which the terms, H_I and $H_{s.r.}$, play metonymic roles. The long answer, which may shed light on the short one, requires us to follow Duhem's lead, and undertake an analysis of *theoretical representation* as it appears in BP III, a project that will take us to the end of Part 2.2 of this essay.

In the BP quartet we may distinguish two layers of theoretical representation. The first is the representation of a metal by a simplified and idealized model, in this case the jellium model, consisting of a gas of electrons moving against a uniform background of positive charge. The second involves a foundational theory. The model is represented in two ways—as a classical system and as a quantum mechanical system;[66] the quantum mechanical treatment is applied in BP III. As we have seen, because the behaviour of a quantum system is governed by its Hamiltonian, to represent it the theoretician needs to know the system's energy. In the jellium model, it

---

[66] The choice of a foundational theory need not involve an extra layer of representation; instead, we can simply distinguish two jellium models at the first level.

comes from three sources: the kinetic energy of the electrons, the Coulomb energy due to the electrostatic repulsion between them, and their self-energy. These physical quantities are *denoted* (at the second level of representation) in the language of quantum mechanics by standard formulae (slightly modified to take into account the effect of the positively charged background). The sum of these formulae, signified by 'H^1' represents the total energy of the model. By a series of mathematical transformations and judicious approximations, Bohm and Pines *demonstrate* that H^1 is effectively equivalent to another cluster of formulae, H_{new}, which can be *interpreted* in physical terms: $H_{electron}$ contains terms referring only to individual electrons; 'field coordinates appear only in H_{coll}, and thus describe a set of uncoupled fields which carry out real independent longitudinal oscillations' (619b); and $H_{res\ part}$ 'describes an extremely weak velocity-dependent electron-electron interaction' (620b).

I have emphasized the words 'denote', 'demonstrate', and 'interpret' because I take the sequence of these actions to be the essential core of the practice of modelling in physics.[67] I call it 'the DDI account of representation'.[68] It applies to theoretical models and physical models alike. The term 'denotation' has the same meaning when applied to either type of model, as does 'interpretation', but the sense attached to 'demonstration' depends on whether a physical or a theoretical model is involved. In the case of physical models, 'demonstration' has its usual meaning; a ripple tank, for example, can be used to *demonstrate* interference patterns. In the case of theoretical models, 'demonstration' is given its seventeenth-century meaning, which corresponds to our present-day 'deduction'. In both cases, however, the demonstration phase of the sequence has the same function. It enables the model to act as an epistemic engine, generating illumination.

The DDI account allows for more than one layer of representation. At the first level of representation, Bohm and Pines represent a metal by the jellium model, denoting the conduction electrons of the metal by the electrons of the model, and the combined charge of the nuclei and valence electrons of the metal by a uniform positive charge. To demonstrate that the electrons of the model exhibit collective behaviour, Bohm and Pines move to the second layer of representation. As they do so they begin both the denotation phase of the second layer of representation and the demonstration phase of the first. To put this another way: the demonstration phase of the first layer

---

[67] See Hughes 1997.   [68] Essay 5 is solely devoted to the DDI account of representation.

of representation is constituted by the entire DDI sequence of the second, the sequence outlined two paragraphs ago, in which the terms 'denote', 'demonstrate', and 'interpret' were first introduced. In one respect, however, that paragraph blurred the distinction between the first layer of representation and the physical world by referring, as do Bohm and Pines, to 'the kinetic energy of the electrons' of the jellium model, and, more generally, to the 'physical quantities' associated with it. Strictly speaking, the 'electrons' of this model are abstract entities, while electrons *tout court* inhabit the physical world. Furthermore, as Cartwright (1999) has pointed out, in a model like the jellium model, not only is the 'environment' of the 'electrons' simplified, but so are the 'electrons' themselves: they have 'charge' and 'mass', but no 'spin'.

Needless to say, Bohm and Pines do not enclose terms of this kind in quotation marks, and with good reason. In the first place, no one would expect to read, in a description of Velázquez' *Las Meninas*, the sentence: 'In the "foreground" a "dog" is "lying", "looking out" at us.'[69] In the second place, whereas in the 'denotation phases' that yielded H^1, Bohm and Pines drew attention to the intermediary role played by the jellium model, in interpreting H_{new} they collapse the two levels of representation and move directly from the three components of H_{new} to the physical world, where quotation marks are not required. The collapse is made explicit in Bohm and Pines' comment on the component H_{coll}: '[O]ur field coordinates occur only in H_{coll}, and thus describe ... *real* longitudinal oscillations' (619b; my emphasis).

The two-stage denotation of the physical systems by H^1 and the interpretation in physical terms of H_{new} mark off the two boundaries of the demonstration phase. Within this phase, however, there is a continuous interplay between the formulae of the second level and the physical quantities they represent. Indeed, it would not be an overstatement to say that this interplay drives the deduction from H^1 to H_{new}. Accompanying the steps that lead from H^1 to H_{new} is a commentary on their physical implications. Only one of these steps is purely formal: Step $(3 \to 4)$ consists solely of the *trivial move*. All the others have physical import. Even in Step (1), when Bohm and Pines write the energy for the Coulomb interactions between electrons as a Fourier series, the choice has two rationales. The announced rationale is that the choice allows them to 'take into account the uniform

---

[69] Magritte's 'Ceci, ce n'est pas une pipe' notwithstanding.

background of positive charge' by the simple device of excluding from the summation over k in the Fourier series the index $k = 0$ (610b). The long-term rationale is that the new idiom is ideal for expressing oscillatory phenomena. The subsidiary conditions on the wave function introduced in Step $(1 \rightarrow 2)$ guarantee that Maxwell's equations are satisfied. Step $(1 \rightarrow 3)$ is prefaced by an argument which justifies by *Physical analogy* the truncation of the Fourier series of H^2. The remaining steps yield the Hamiltonians H^4, H^5, and H_{new}. Each of these—not only H_{new}—is broken down into clauses that are given physical interpretations. H^4 and H^5 share three clauses: $H_{\text{part}}, H_{\text{I}}$, and $H_{\text{s.r.}}$, and a fourth clause of H^4 ($U^\dagger$) is discarded on both theoretical and physical grounds (*Perturbation theory* and the random phase approximation). (The replacement of the last clause in H^4 (H_{osc}) by H_{field} in H^5 has no physical significance; it relies on the second *Strictly mathematical move*.) Step $(5 \rightarrow \text{new})$ involves not only the canonical transformation of H^5 into H_{new} (an example of the third *Strictly mathematical move*) but the assumption that a particular equation gives us the *dispersion relation* (the relation between the frequencies of the collective oscillations and their wavelengths, justified by *Physical analogy*).

The clauses H_{I} and $H_{\text{s.r.}}$ in H^4 are, of course, the formulae which were endowed with causal powers in the second *Causal inference*. We are now in a position to make sense of that endowment. Bohm and Pines interpret H_{I} as 'a simple interaction between the electrons and the collective field' (612b), and $H_{\text{s.r.}}$ as 'the short-range part of the Coulomb interaction between the electrons' (ibid.). We may reverse the metonymy that the second *Causal inference* relied on, and replace H_{I} and $H_{\text{s.r.}}$ by their interpretations to obtain a straightforwardly causal argument.

From [the expressions for the components of H^5] we see that if we neglect *a simple interaction between the electrons and the collective field*, the collective oscillations are not affected at all by '*the short-range part of the Coulomb interaction between the electrons*'. Thus, '*this short-range part of the Coulomb interaction*' can influence the q_k only indirectly through *the interaction between the electrons and the collective field*. But, as we shall see, the *direct* effects of *the interaction between the electrons and the collective field* on the collective oscillations are small. Thus, it may be expected that the *indirect* effects of '*the short-range part of the Coulomb interactions*' on the q_k through *the interaction between the electrons and the collective field* are an order of magnitude smaller and may be neglected in our treatment which is aimed at approximating the effect of *a simple interaction between the electrons and the collective field.* Notice that the argument is now couched in the narrative mode illustrated in section 2.2.4.

The first topic of this section was the vending machine account of deduction; the last one the role of representation in theoretical physics. In Nagel's version of the vending machine, the input and the output are statements, and the mechanisms which transform one into the other are linguistic. Its components are statements, and its workings are governed by the laws of logic. If the machine is to work, then although the referring terms are divided into two disjoint classes, observational terms and theoretical terms, and correspondence rules are needed to link the two, all these elements must belong to a single language. On Duhem's account of the structure of physical theory, however, the terms of our theories bear to actual physical systems 'only the relation of sign to thing signified' (*PT*, 20). On a first reading there is no conflict here. Nagel is talking simply about a language, and Duhem is talking about the relation between language and the world it describes; in any language, the relation between word and object is that of the sign to the thing signified. But, as we read further, it becomes clear that the 'signs' Duhem has in mind are elements of a strictly theoretical language, and the 'things signified' are parts of the physical world. He distinguishes between 'theoretical facts' and 'practical facts'.

In … a theoretical fact there is nothing vague or indecisive. Everything is determined in a precise manner: the body studied is geometrically defined; its sides are true lines without thickness, its points true points without dimensions … Opposite the *theoretical* fact let us place the *practical* fact translated by it. Here we no longer see anything of the precision we have just ascertained. The body is no longer a geometrical solid; it is a concrete block. However sharp its edges, none is a geometrical intersection of two surfaces; instead, these edges are more or less rounded and dented spines. Its points are more or less worn down and blunt.   (ibid. 132–4; emphases in the original)

The gap between theoretical and practical facts rules out the possibility of logical relations existing between sentences expressing different types of fact, and that in turn rules out a strictly deductive account of theoretical explanation.[70]

In the passage above, Duhem anticipates the insight that underlies much of the philosophy of physics in the last two decades, the insight that theories

---

[70] The difficulties in carrying through the project are acknowledged by Nagel in chap. 11 of *The Structure of Science* and very clearly by Carl Hempel in his essay, 'The Theoretician's Dilemma' (1965).

provide accurate descriptions, not of actual physical systems, but of the simplified and idealized simulacra that we call 'models'.[71] On the other hand, Duhem did not anticipate developments of physics itself. As a card-carrying energeticist, he had no time for 'the atomic hypothesis'. In contrast, forty-five years after the publication of his *The Aim and Structure of Physical Theory*, Bohm, Pines, and their contemporaries took it for granted that, impurities aside, a metal in the solid state consisted of crystals, each comprising a regular lattice of ions surrounded by a gas of electrons. And, given this description, the physicist faced a true 'Theoretician's Dilemma'. In Pines' words (1987: 68), which I quoted very early on in this part of the essay: 'In any approach to understanding the behaviour of complex systems, the theorist must begin by choosing a simple, yet realistic model for the behaviour of the system in which he is interested.'

The dilemma resides in the phrase 'a simple, yet realistic model'. For verisimilitude begets complexity, and complexity begets intractability. The dilemma is not new; the 'many-body problem' was faced by Newton, for whom 'many' was 'three'. He supplemented the deductive resources of Apollonian geometry with the first perturbation techniques of the modern age. (Ptolemy's epicycles, equants, and eccentrics were his counterparts in the ancient world.) Today it is neither surprising nor reprehensible that physicists should supplement the resources of strict deduction, and that these supplements should be adopted by others. The random phase approximation, for example, is arguably the most valuable contribution to the deductive practices of solid-state physics directly traceable to Bohm and Pines.

Though they arrived too late for Bohm and Pines to make use of them, two other supplements that appeared around 1950 should be noted. One emerged within physics, the other from elsewhere. In 1949, Feynman introduced what were soon referred to as 'Feynman diagrams', and by the mid-1950s they were part of the toolkit of many theoretical physicists—including those working in solid-state physics.[72] And, as early as 1956, a paper on solid-state physics by Neville Mott (1956b: 1205) began as follows: 'The use of electronic

---

[71] Indeed, without changing its content, we could rewrite Chapter III of Part 2 of *The Aim and Structure of Physical Theory*, in which the distinction between theoretical and practical facts is made, in terms of 'models', as I use the term. Duhem himself reserved 'model' for those imaginary mechanical models by which nineteenth-century English physicists, with their 'ample, but weak minds' (ibid. 63) sought to explain electromagnetic phenomena.

[72] I say more about Feynman diagrams in Section 2.3.1 of this essay. We may note that in 1967 Richard Mattuck published *A Guide to Feynman Diagrams in the Many Body Problem*, a book that was entirely devoted to applications of Feynman diagrams.

computers has made it possible to calculate the electronic wave functions of simple molecules with any degree of accuracy desired.'

From our present perspective we can hardly regard the use of computer techniques as a 'supplement' to deductive practice. Rather, virtually all the work performed in the demonstration phases of theoretical representation is now outsourced to computer programs. Paradoxically, the greatest change of theoretical practice in the last half-century did not involve any change of theory.

2.3. ON THEORETICAL PRACTICE

2.3.1. Theoretical Practice and the Bohm–Pines Quartet

My chief aim in this essay has been to provide a case study of the theoretical practices of physics. I chose the quartet of papers that David Bohm and David Pines published in the early 1950s, and in this section of the essay I will use this example to summarize what *theoretical practice* involves, and to sketch the ways in which this practice changes though time. Initially, this will involve some recapitulation of familiar material, but some new themes will be sounded as I go on.

For most readers of the BP quartet, its importance lay in its third and fourth papers, in which the authors used the resources of orthodox quantum mechanics to investigate the behaviour of the conduction electrons in metals.[73] The mathematical foundations of that theory had been laid twenty years earlier by Paul Dirac and John von Neumann; Dirac's *The Principles of Quantum Mechanics* was published in 1930 and von Neumann's *Mathematische Grundlagen der Quantenmechanik* in 1932. The greater part of Dirac's work presented an abstract formalism of which both Heisenberg's matrix mechanics and Schrödinger's wave mechanics (published in 1925 and 1926, respectively) were realizations;[74] von Neumann's work absorbs Dirac's 'transformation theory' within the mathematics of 'Hilbert spaces'. (Both terms were coined by von Neumann.) Each of these publications marked a major advance in mathematical physics, but I mention them only to emphasize that, on two counts, none of them provided an

---

[73] The exception was Stanley Raimes, whose 1961 book, aimed at students of experimental physics, presented the classical account of plasma oscillations given in BP II.

[74] Schrödinger had previously (1926) shown the equivalence of the two approaches.

example of *theoretical practice*, as I use the phrase. The first is that, even though a progression can be traced from Heisenberg to von Neumann, all four were too original to be regarded as part of an established practice; rather, they were the innovations around which practices would coalesce. The second is that theoretical practices emerge when a theory is put to work—that is to say, when a mathematical theory is applied to a particular phenomenon or system, and must be supplemented by additional procedures and techniques.

Twenty years separated the publication of the Bohm–Pines quartet from the work of Dirac and von Neumann. In that period, not only did physicists become thoroughly familiar with the intricacies of quantum mechanics, but they also devised a variety of stratagems and artifices to complement the bare mathematics of the theory. Bohm and Pines in their turn contributed a number of original techniques, as we shall see, but it is the *unoriginal* part of their work, the miscellany of theoretical elements which they inherited from previous investigators, and with which they and their peers were fully conversant, that allows us to speak of a 'theoretical practice' shared by a community of practitioners.[75]

A partial inventory will show just how diverse the elements of this practice were. From the mathematical part of orthodox quantum theory, Bohm and Pines took standard results; some came from the algebra of Hilbert space operators, others from the standard perturbation techniques of the theory. On the physical side, the basic structure of the system they investigated was well known. Metals in the solid state were crystalline; within each crystal, the positive ions of the metal formed a regular array, the 'crystal lattice', and were surrounded by a cloud of valence electrons. To apply quantum mechanics to this system, Bohm and Pines needed to model it in a way that adequately satisfied the competing desiderata of verisimilitude and simplicity. The criteria for the latter rested on the exigencies of quantum theory—in particular, on how amenable the model was to representation by the stock Hamiltonians of quantum mechanics.[76] In BP III, by choosing a model in which the discrete nature of the individual ions was ignored, the authors were enabled to write down a comparatively simple Hamiltonian for the

---

[75] The emphasis on a community of practitioners immediately raises the question: Is it a solecism to speak of 'Newton's theoretical practice'? It is not. But, Newton is a unique case. No other theoretical physicist has simultaneously articulated a truly original theory and applied it to as many diverse phenomena as did Newton.

[76] This point has been emphasized by Cartwright (1999: 268–78) in analysing the Hamiltonian used by Bardeen, Cooper, and Schrieffer in their theory of superconductivity.

system, H^1. It contained just three terms. Each was a standard expression for a particular component of the energy of an aggregate of electrons, slightly modified to allow for a background of positive charge. The term for the Coulomb potential was expressed as a Fourier series, and when H^1 was rewritten in terms of the longitudinal vector potential $\mathbf{A(x)}$ and the electric field intensity $\mathbf{E(x)}$ of the electromagnetic field within the plasma, each of them was also decomposed by Fourier analysis into a series of sinusoidal waves. Both moves were well established. Indeed, the latter was the most venerable procedure used in the quartet: Jean Baptiste Fourier was born in 1768, a year before Napoleon Bonaparte.

Again, of the four kinds of approximations which Bohm and Pines drew attention to in BP I (see section 2.2.5), only the fourth, the random-phase approximation, was original. The first approximation allowed electron–electron collisions and electron–ion collisions to be ignored. The abstraction which ignored electron–electron collisions was analogous to that which ignored atom–atom collisions in a simplified nineteenth-century derivation of the ideal gas law, and (as I pointed out in section 2.2.5) individual electron–ion collisions were effectively ignored by all theoreticians who used the jellium model. The second approximation relied on the assumption that the organized oscillations in the plasma were small, in order that the authors could use 'the *customary linear approximation* appropriate for small oscillations' (my emphasis); the third, the neglect of terms involving v^2/c^2 on the assumption that the velocity v of the electron is small compared with the speed of light, c, was hardly a novel manoeuvre (see section 2.2.7).

The elements of theoretical practice are often easily identified; their use is so widespread that they have acquired names. The *jellium model* is an example. In that instance, the name is descriptive of the element, but more often the name attached to an element is the name of its originator, as in 'a Fourier series'. In BP III alone we find eleven such names attached to thirteen elements of five different kinds: to physical phenomena [*Coulomb interactions* (609a, *passim*)]; to theoretical models [*a Fermi gas* (610b), *a Bose field* (625a)]; to magnitudes associated with such models [*the Debye length* (611b), *the Bohr radius* (615a)]; to mathematical entities [*a Fourier series* (610b), *the nth Hermite polynomial* (613a), *the Slater determinant* (613a), *the Fermi distribution* (615a)]; and to mathematical representations of physical systems [*Maxwell's equations* (611a), *the Heisenberg representation* (623b), *Fermi statistics* (623b), and, of course, *the Hamiltonian*, which appears on virtually every page]. Two of these phrases are used in describing the kind of theoretical approach taken by others; the rest are woven into the authors' own arguments. There are no footnotes to provide glosses on them, nor are they needed. The phrases are part of the vernacular of the readers of the *Physical Review*, and the elements they denote are part of a physicist's stock-in-trade, not least because of their versatility. The 'Debye length', for example, entered the physicist's lexicon in 1926, as the thickness of the ion sheath that surrounds a large charged particle in a highly ionized electrolyte. As we have seen, twenty-five years later, Bohm and Pines used it in their treatment of conduction electrons in a metal, at a scale several orders of magnitude smaller than the phenomenon it was originally designed to model. Because of this versatility, when Bohm and Pines wrote in the Introduction to BP II, 'For wavelengths greater than a certain length λ_D the fluctuations are primarily collective', this parenthetical aside conveyed two messages. It gave the reader information, and it told him that he was on familiar ground.

So much for the techniques that Bohm and Pines inherited. In turn, the quartet bequeathed a number of useful theoretical tools to future investigators. The most obvious is the 'random phase approximation'. Another is the practice of treating a system consisting of a particle and its immediate environment as a *quasi-particle*.[77] In BP III, the authors observed that, when an electron is surrounded by a cloud of collective oscillations, it behaves as though its mass has increased, and they treated it accordingly. (See the penultimate paragraph of section 2.1.4.) Subsequent examples of quasi-particles include (a) a *conduction electron*, which consists of an electron moving within the periodic potential provided by fixed lattice ions; (b) a *polaron*, which is an electron moving in an insulating polar crystal; and (c) a *quasi-nucleon*, which is a proton or a neutron surrounded by a cloud of other nucleons.[78] Another bequest was made by Pines in 1956, when he proposed that the quantum of energy lost by a high-energy electron when it is scattered within a metal foil should be regarded as a particle: a *plasmon*. He summarized 'the evidence, both experimental and theoretical, which points to the plasmon as a well-defined entity in nearly all solids' (1956: 184b). Plasmons are analogous to phonons. Both are quanta of energy in condensed matter, but, while

---

[77] Nearly a decade later, Pines included a section (1962: 31–4) on the definition of quasi-particle in the lecture notes that form the first part of *The Many-Body Problem*. The remainder of the volume is an anthology of papers from previous years on the physics of many-body problems.
[78] See Mattuck 1967: 15.

phonons are stored as thermal energy by atoms as they vibrate about their mean positions, plasmons are stored by the valence electrons of a metal either individually, or collectively in plasma oscillations.

But, arguably the most significant legacy of Bohm and Pines' work was the contribution it made to the theory of superconductivity published by Bardeen, Cooper, and Schrieffer (BCS) in 1957. The overall strategy of their paper resembled that used in BP III. To echo Ziman, in both papers a model was set up, its Hamiltonian was prescribed and its theoretical properties deduced, and experimental phenomena were thereby explained. The BCS model was the more complicated of the two. Whereas the BP model represented the charge of the positive ions as a uniform background of charge, the BCS model not only allowed for the fact that the ions formed a regular lattice, but took into account the lattice vibrations as well. These vibrations (and here I switch into narrative mode) are quantized into phonons, which mediate an attractive interaction between two valence electrons (the so-called 'Cooper pairs'). Since the repulsive force of the Coulomb interaction between these electrons is attenuated by the screening effect predicted by Bohm and Pines, the net force between the members of a Cooper pair may be attractive, causing the metal to become a superconductor.

The Hamiltonian for the BCS model reflects these two opposing interactions. It contains four terms (Bardeen, Cooper, and Schrieffer 1957: 1179a), of which the third, H_{coul}, represents the energy due to the screened Coulomb interactions between electrons, and the fourth (oddly referred to as H_2) the energy due to the phonon interaction with the Cooper pairs. For an expression for H_{coul} the authors go to the BP theory. Recall from sections 2.1.4 and 2.1.5 that what had started out in BP III as the Hamiltonian H^1 for a dense electron gas was successively modified until it appeared in P IV in its final version: $H = H_{part} + H_{coll} + H_{s.r.}$. In that final Hamiltonian, the effects of Coulomb interactions were divided into two parts. The long-range effects were summarized in the term H_{coll}, where they were 'effectively redescribed in terms of collective oscillations of the system as a whole' (P IV: 627a). In contrast, the term $H_{s.r.}$ corresponded to 'a collection of individual electrons interacting with a comparatively short-range force' (ibid.). Since the high energies associated with collective oscillations do not occur in a superconductor, the result is that the term H_{coul} in the BCS Hamiltonian need only represent short-range effects—that is, the screened interaction represented by $H_{s.r.}$.

The kind of theoretical practice exemplified by the BP theory of plasmas and the BCS theory of superconductivity did not last. From the perspective of the early twenty-first century, these theories appear as two of the last constructionist contributions to solid-state physics. I borrow the term 'constructionist' from Philip Anderson (1972), who contrasts two hypotheses: constructionist and reductionist. The latter is a hypothesis about the physical world. On the reductionist hypothesis, the behaviour of a macroscopic system is ultimately determined by the behaviour of its submicroscopic constituents, which in turn is governed by simple fundamental laws. The constructionist hypothesis is bolder; it is a hypothesis about the reach eventually attainable by our fundamental theories of physics. It suggests that, as and when our 'final theory' has established the ontology of fundamental particles and the laws which they obey, we will have the theoretical resources to explain all the phenomena of nature. This hypothesis Anderson rejects: '[T]he reductionist hypothesis does not by any means imply a "constructionist" one: the ability to reduce everything to simple fundamental laws does not imply the ability to start from those laws and reconstruct the universe' (ibid. 393). To amplify this, he continues, '[I]t seems to me that we may array the sciences linearly according to the idea: The elementary entities of Science X obey the laws of Y', and he then sketches a hierarchy whose first three entries are these:

X	Y
solid-state or many-body physics	elementary particle physics
chemistry	many-body physics
molecular biology	chemistry

And he quickly adds, 'But this hierarchy does not imply that science X is "just applied Y". At each stage, entirely new laws, concepts and generalizations are necessary …'.

I extend the usage of the adjective 'constructionist' to cover theoretical endeavours that would, if successful, confirm the constructionist hypothesis. In this sense, the approach taken by Bohm and Pines was constructionist in many ways. This is beautifully illustrated by the transformations that I alluded to two paragraphs ago, whereby the Hamiltonian H^1 in BP III became H in P IV. Bohm and Pines started with elementary particles; the components of H^1 dealt with the properties and interactions of electrons: their individual kinetic energies, their individual self-energies, and the energy due to the pairwise Coulomb forces between them. As we saw, the final Hamiltonian is the sum of three terms, $H_{part} + H_{coll} + H_{s.r.}$. The first and third terms still deal only with electrons. H_{part} represents their kinetic energy and self-energy (albeit as those were modified by the environment) and $H_{s.r.}$ 'corresponds to a collection of individual electrons interacting via a comparatively weak short-range force' (P IV: 627b). H_{coll}, however, is expressed entirely in collective variables: 'The long-range part of the Coulomb interaction has effectively been redescribed in terms of the collective oscillations of the system as a whole' (P IV: 627a). In this way, the emergent behaviour of the system is made explicit, and the microscopic and the macroscopic are accommodated under one roof.[79]

But, the collective variable approach to many-body problems had severe limitations. It was never taken up outside David Bohm's immediate circle at Princeton. Even the authors of the BCS paper had reservations about its use; in the conclusion of their paper, they wrote (Bardeen, Cooper, and Schrieffer 1957: 1198a): 'An improvement of the general formulation of the theory is desirable', and listed half a dozen items which could be improved on, the BP 'collective model' being one. From the middle of the 1950s on, attention moved away from constructionist treatments of high-density plasmas in favour of macroscopic treatments. For example, in 1954, Lindhard, a Danish physicist, described the behaviour of plasma entirely in terms of a macroscopic property, its dielectric constant, and his example was quickly followed by others. Intertwined with the rejection of a constructionist methodology was a major change in theoretical practice: the orthodox quantum mechanics used by Bohm and Pines was supplanted by quantum field theory.[80]

Throughout the 1930s and 1940s, quantum field theory had been highly problematic. It had a number of successes, like the prediction of the positron, but it was prone to divergencies—that is to say, seemingly innocuous calculations went to infinity.[81] In 1949, the problem was resolved from two directions. Sin-itiro Tomonaga and Julian Schwinger used a generalization of operator methods, which was theoretically impeccable but very difficult to work with; Richard Feynman, on the other hand, used a 'propagator

[79] Subsequently, Pines predicted the value of an emergent property of the plasma, its specific heat (P IV: 632).

[80] Nevertheless, despite the change of methodology and of foundational theory, the various approaches to high-density plasmas in the 1950s were by no means incommensurable. For instance, Nozières and Pines bridged the gap between the collective variable approach and the alternatives in papers like 'Electron Interactions' (1958b), which was subtitled, 'Collective Approach to the Dielectric Constant'.

[81] For a succinct account of the theory's successes and failures, see Howard Georgi 1989: 449. The seriousness of the problem can be judged by the language used to describe it. Michio Kaku (1993: 4) describes quantum field theory as 'plagued with infinities', and for Georgi that was its 'tragic flaw'.

approach', and showed how it could be pictorially represented by simple diagrams.[82] While the Tomonaga–Schwinger approach made quantum field theory respectable, the Feynman approach made it easy to apply. Later in the year, Freeman Dyson showed the two approaches to be equivalent and by 1955 quantum field theory had established itself as the core of a new theoretical practice, largely because the Feynman diagram had shown itself to be one of the most remarkable theoretical tools of the twentieth century. In particular, there were two major reasons why physicists found the theory well suited for treating dense electron gases. The first was '[t]he realization, that there exists a great formal similarity between the quantum theory of a large number of Fermi particles and quantum field theory' (Hugenholtz and Pines 1959: 489/332); the second was the theoretical economy afforded by Feynman diagrams, and was as important as the first.[83] Eugene Gross, who, like Pines, was one of Bohm's students and had used the collective variables approach, writes (1987: 47): 'Feynman's introduction of diagrams freed the imagination of theoretical physicists to deal with what had been depressingly complicated formalisms in quantum field theory and many-body physics.'

2.3.2. On Methodology: A Very Brief Note

To my mind, there are at least two respectable ways in which a philosopher of physics can approach theoretical physics. The first is to examine specific theories with an eye to philosophical issues. Examples are the usual suspects: statistical mechanics, general relativity, and quantum theory. But the theories investigated need neither have an extensive domain of application nor be widely accepted; Bohmian quantum mechanics and Heinrich Hertz's version of classical mechanics are cases in point. Some of the issues will be metaphysical: *the problem of 'the direction of time'; what kind of being does space-time have?* Others will be internal to the theory: *what is measurement in quantum theory?*[84] Though a theory may be associated with a particular physicist, as Einstein is with the special and general theories of relativity, that

---

[82] Not all physicists welcomed the Feynman diagram. In fact, Schwinger is reputed to have forbidden his graduate students from using it, on the grounds that its use represented the triumph of theft over honest toil.

[83] In section 2.3.1, where I discussed alternative approaches to electron plasmas between 1954 and 1958, I cited fourteen papers from that period. Feynman diagrams appear in nine of them.

[84] These three issues are section headings in Lawrence Sklar's admirable *Philosophy of Physics* (1992).

is irrelevant to the issues the theory raises for philosophers. A theory achieves a life of its own, so to speak.

The second approach consists in examining how physicists use these theories, and is more empirical in nature. Its object is, first of all, to give accurate descriptions of theoretical practices, something signally absent in philosophical circles until quite recently. On this approach, the philosopher regards each specific application of the theory as uniquely tied to the theorist (or theorists) who made it. It may involve standard techniques, like the use of the Bloch Hamiltonian or the random phase approximation, but those are choices the theorist opts for. The material examined by the philosopher is the written word, and its genre the monograph or the paper in a physics journal. To amplify a theme from the Preamble, such a philosopher regards such works as texts, and his task to be analogous to the literary critic's. Like a good critic, the philosopher using this methodology draws the reader's attention to the elements of a text and how they fit together, while recognizing the stern constraints the theorist labours under. In short, he or she shows the reader what kind of text it is, and the nature of its success.

This essay is an endeavour of the second kind.

3

Laws of Physics, the Representational Account of Theories, and Newton's *Principia*

I showed what the laws of nature were, and tried ... to show that they
are such that, even if God created many worlds, there could not be any
in which they fail to be observed.

René Descartes[1]

The 'laws of nature' ... are our own free creations; our inventions.

Karl Popper[2]

PREAMBLE

In Chapter 8 of Book II of the *Physics*, Aristotle argues for a teleological
account of nature.[3] His argument has the form of a disjunctive syllogism:
Things happen by chance or to some end; it is implausible that the regularities
in nature happen by chance; therefore they happen to some end. The same
syllogism reappears in Thomas Aquinas' Fifth Way,[4] and, as late as the last
decade of the sixteenth century, it informs the debate between Mutabilitie
and Nature in the final stanzas of Edmund Spenser's *The Faerie Queene*.
But, half a century later, Descartes declares, in Part I of the *Principles of
Philosophy*, 'We shall entirely reject from our Philosophy the search for final
causes'.[5] Aristotle's syllogism does not hold him in its sway, because he
rejects its major premiss. For Descartes, the alternatives it presents—that
things happen either by chance or by some end—are not exhaustive. A third
possibility is that things happen as they do in conformity with the laws of

---

[1] René Descartes, *Discourse on the Method*, Pt 5 ([1637] 2001): 132. Descartes is here outlining
the strategy he had pursued in *Le Monde*, a work which he withdrew from publication, but was
published posthumously. [2] Popper 1959: 79.

[3] Aristotle, *Physics* II.8: 199a.   [4] Aquinas, *Summa Theologica*, I, Q. II, Art. 3.

[5] Descartes [1644] 1983a: 14.

nature, laws that natural processes cannot disobey, laws that define physical necessity.

Descartes identified the laws of nature with the laws of motion.[6] Half a century later, Newton presented an alternative set of laws of motion as Axioms in his *Principia* (1686), and I will return to these laws in Part 3.5 of this essay. Not surprisingly, the rational mechanics of the eighteenth century was articulated in terms of laws; in fact, Euler and d'Alembert each declared that the laws associated with them (in Euler's case, his law of dynamics, in d'Alembert's, his laws of equilibrium) were necessary, not contingent, truths, an issue that also exercised the Academy of Berlin:[7] 'Why is it more than probable, that all men die; that lead cannot, of itself, remain suspended in the air; that fire consumes wood and is extinguished by water; unless it be that these events are found agreeable to the laws of nature' ([1748] 1975: 114–15).

Up to 1750, however, mechanics was almost the only physical science which boasted laws. The catalogue of subdivisions of 'Science of Nature' in d'Alembert's *Preliminary Discourse to the Encyclopedia of Diderot* included one that identified the various types of 'mixed (i.e. applied) mathematics'. This category contained the two branches of mechanics: '*statics* and *dynamics*'; '*geometric astronomy*, whence comes *cosmography*'; '*opticks*'; '*acoustics*'; '*pneumatics*'; and '*the art of conjecture*, whence is born the *analysis of games of chance*'. The only laws connected with these sciences that d'Alembert mentions in the *Discourse* are those dealing with mechanics.[8]

Needless to say, between 1650 and 1750, there were scientific advances that were subsequently described in terms of laws. Robert Boyle, for instance, never referred to the relation he had discovered between the pressure and volume of air as a *law*, nor did he present his finding in a way that readily lent itself to the succinct form of a physical law.[9] More curious is the case of Newton's *Opticks*, first published in 1704. Although the work begins, as does the *Principia*, with a chapter of Definitions, followed by another introducing Axioms, only in the *Principia* are the axioms referred to as 'laws'. In fact, the word 'law' never occurs in that edition of the *Opticks*. In Query 31 of the second English edition (1717), however, Newton proposed

[6] See §23 of Part II of Descartes [1637] 2001: 50. Note that, for Descartes, the motion of a body was proportional to the product of its speed and its extension—i.e. its volume, and the speed was not a vector quantity.   [7] See Dugas 1988: 242, 247.

[8] They are 'the laws of equilibrium and movement' (d'Alembert [1751] 1963: 21): 'laws of the descent [of falling bodies] down inclined planes', and 'the laws of the moments of pendulums' (ibid. 24), 'laws of percussion' (ibid. 27), and 'laws of mechanics' (ibid. 82).   [9] See Shapin 1994: 7.

that the search for laws should be part of the methodology of (at least one part of) natural philosophy. In a discussion of 'attractive Powers', he writes ([1726] 1934: 376): '[W]e must learn from the Phenomena of Nature what Bodies attract one another, and what are the Laws and Properties of the Attractions.' The Attractions cited were not only gravitational attraction but also the attractions of magnetism and electricity (which we now label 'electrostatics').

Of these Attractions, Newton knew that the 'Attraction of Gravity' obeys an inverse square law of the distance between bodies with weight; but, as regards the other two Attractions, all he knew was that they reached 'to very sensible distances'. That vague estimate had to wait nearly eighty years, until Charles Augustus Coulomb used his torsion balance to show inverse square laws for each of them.[10] And, by the end of the eighteenth century, other parts of physics (as we call it) were becoming respectable. In 1795, Christoph Heinrich Pfaff wrote:

Through the determination of laws, the overview of various phenomena is facilitated, heterogeneity is happily transformed into a harmonious way of looking at things, and we are always more prepared to investigate the causes, since the constant coexistence of changes in the external circumstances suggests a mutual causal relationship. (Newton: [1686] 1934)

Pfaff wrote in the wake of Galvani's experiments,[11] and the idiom he uses ('the overview of various phenomena', for example) shows that the scope of physical laws was little further than mechanics.

Until about 1900, the lengthening list of laws of physics bespoke the development of physics as a science. The nineteenth century in particular saw the proclamation of Ampère's Law, Avogadro's Law, Coulomb's Law, Graham's Law, the Ideal Gas Law, Kirchoff's Law, Ohm's Law, Stefan's Law, Stokes's Law, and Wien's Law, to name but a few. In 1905, Pierre

---

[10] In the cases of magnetic poles and electrostatic charges, there are two forces in each: attractive in the case of two unlike poles or charges; repulsive in the case of like poles or charges.

[11] Here Pfaff is anticipating a series of experiments based on earlier experiments by Luigi Galvani in 1791 which caused a multitude of frogs to come to a shuddering end. Those experiments involved frogs on a plate and a charged metal probe. Alessandro Volta used a similar apparatus in 1793, and the physical results were the same, but, for the two experimenters, what was going on was very different. Galvani, who was a physician, located the cause to be the excitation in the frogs' muscles, while Volta, who was a physicist, located the cause in the charged probe. Pfaff's enterprise changed the focus again: his experiments focused on pairs of metal probes, while the frog's muscles were just a conduit for the electric charges to pass from one probe to another. More information can be found in Trumpler 1997.

Duhem, that staunch proponent of energetics, declared that the aim of a theory was 'to represent as simply, as completely, and as exactly as possible a set of experimental laws' (*PT*, 19). Almost three decades later, in 1942, Carl Hempel wrote (1965: 231): 'The main function of laws in the natural sciences is to connect events in patterns which are usually referred to as *explanation* and *prediction*.'

Late twentieth-century physics gave us few laws, but was none the worse for that. Consider *The New Physics* (Davies 1989), a collection of eighteen essays on post-1970 developments in physics, with 516 pages of text, fifty-eight pages of glossary, and twelve pages of index. That index contains only eight entries referring to *laws*. The laws cited are (a) Kepler's Laws (from the early seventeenth century); (b) the inverse square law for light (also first enunciated by Kepler); (c) Hubble's Law (1927), that 'roughly linear relation between velocities and distance' of spiral nebulae, and called by astronomers the 'red shift';[12] (d) a generic entry under 'laws of physics'; and (e) just two laws from the physics of the second half of the twentieth century. The first, from Essay 2, is a topological conservation law, which supplied the four laws of black hole mechanics (Davies 1989: 29; Will 1989);[13] the second is a law governing a system of quarks: '[T]he baryon of a system is defined as the total number of quarks which it contains, minus the number of antiquarks, all divided by three' (ibid. 40; Guth and Steinhardt).

The paucity of laws in *The New Physics*, I have pointed out. I will discuss conservation laws in Part 3.3 of this essay, and the critical point laws in Essay 6. Again, if physicists no longer seek experimental laws, but nevertheless produce successful theories, physical theory is not what Duhem took it to be. In Part 3.2, I set out an alternative view, which I call the representational account of theories. Nor is the scope of the account limited to the physics of the twentieth century, as I show in Part 3.4, by taking Books I and III of Newton's *Principia* as a case study and examining them from the perspective that the representational account affords. I argue in Essay 7 of this book that this account provides an alternative to Hempel's model of explanation, and in Part 3.2 of the present essay I apply it to an analysis of laws.

I postpone until Part 3.3 of this essay the question whose answer Descartes and Newton took for granted: Are the Laws of Physics the Laws of Nature?

---

[12] For more information, look to Steven Weinberg's *Gravitation, and Cosmology* (1972), 417–18.

[13] 'These laws are in exact parallel with thermodynamics, with κ playing the role temperature, and the areas playing this role of the entropy.'

3.1. THREE PROBLEMS

Despite the emphasis that scientists and philosophers of science have placed upon laws, their nature has remained obstinately problematic. By way of an introduction, I will outline three problems that any adequate analysis, whether of laws of nature or laws of physics, must deal with. They are the problems of the modal force of laws, the problem of accidental generalizations, and the problem of *ceteris paribus* clauses. All three are familiar.

3.1.1. Modal Force

To borrow a title from the late Thelonius Monk, the first may be called the problem of *Why We Must Obey the Facts*. We talk both of physical necessity and of physical laws. Assuming we are not misguided in so doing, what is the connection between the two?[14]

One alternative is to regard every law as itself a modal statement, asserting that a particular generalization holds true in every physically possible world. But how, then, is the ascription of lawhood to be justified? Clearly, not on the basis of empirical evidence alone. For, ignoring the usual problem of induction, let us assume that we have concluded from our observations that all Fs are Gs. Then no additional observations will justify the inference that all Fs are Gs by physical necessity, or, in other words, that all Fs are Gs in every physically possible world, since the only world about which observations provide information is our own.[15]

Appeal to empirical confirmation, however, is not the only mode of justification that has been suggested. David Armstrong, for instance, who ascribes to laws a 'nomic necessity', justifies this ascription by appealing to its explanatory power. Without a modal concept of law, he claims, 'inductive scepticism is inevitable' (Armstrong 1993: 52), and, since the sceptic's position combines a theoretical denial of our fundamental beliefs with a practical reliance on them, '[I]t is … a most serious objection to a philosophical theory if it has inductive scepticism as a consequence' (ibid. 54). Conversely (we may assume) if a modal account of laws shows that

[14] I will put aside the question whether the necessity of laws is contingent, the result perhaps of a fortuitous or, as some think, providential circumstance in the early development of the cosmos.

[15] This point was made in similar terms by Karl Popper (1959: 433), who nevertheless took laws to express physical necessities.

our inductive practices are rational, that will count in its favour. Armstrong has earlier suggested that if we have observed a regularity (e.g. that all Fs are Gs) then 'one possible explanation [of this regularity] is that it is a law that all Fs are Gs' (ibid. 40). For an appeal to law to constitute an explanation, he goes on, laws must be more than statements of regularities, since 'trying to explain why all observed Fs are Gs by postulating that all Fs are Gs is a case of trying to explain something to a state of affairs part of which is the thing to be explained. But a fact cannot be used to explain itself' (ibid.).

This analysis is then applied to the problem of induction. Armstrong writes:

I hold, however, that a law involves an extra thing, some further state of affairs. The presence of the extra thing ... serves, first, to explain why all the observed Fs are Gs, and, second, to entail that any unobserved Fs there are will be Gs. The postulation of this extra thing is a case of inference to the best explanation. It is rational to postulate what best explains the phenomena. Induction is thus rational, because it is a case of inference to the best explanation.   (ibid. 55)

In chapter 6 of his book, Armstrong goes on to argue that the 'extra thing' that laws tacitly invoke is a relation of necessity between universals. Immediately after this assertion he remarks, 'If such a law really holds, then the explanation will be quite a good one' (ibid.). But, if no such relation exists between the universals F and G, but all Fs are, as a matter of fact, Gs, what then? And, how could one tell? The assertion itself is put forward without argument, as a fact that compels intuitive assent. Yet, there is little reason to accept it. Certainly, it does not hold for the regularities that are our present concern. We don't explain why the straight line joining a planet to the Sun always sweeps out areas in equal times by reminding our interlocutor that this accords with one of Kepler's laws. Instead, we show how a corresponding regularity can be derived within Newtonian theory. Nor does the lawhood of the fundamental postulates of that theory explain their content. To announce, for example, that it is a law of nature that a material will continue in a state of uniform rectilinear motion is not to explain that behaviour; it is to say that such behaviour, being natural, requires no explanation. What need to be explained are departures from changes in the object's speed or direction of motion.[16]

---

[16] Note that the ten-line paragraph that Newton appends to his statement of the first law in the *Principia* is entirely taken up with explaining apparent violations of the law, like the motions of projectiles and of planets (Newton [1726] 1934: 13).

Despite the importance that Armstrong attaches to the explanatory power lawhood confers, he offers no account of explanations, nor of the ways in which laws might figure in them.[17] His appeal to lawhood *tout court* emerges as an appeal to a sort of 'nomic virtue', analogous—perhaps in more way than one—to the 'dormitive virtue' lampooned by Molière in *Le Malade Imaginaire*.

3.1.2. Accidental Generalizations

A more-promising strategy might be to regard physical laws not as inherently modal statements but as a privileged class of truths about the actual world. Once we have specified these truths, then a definition of physical possibility is available to us: a physically possible world is one in which all these truths hold. On this account, laws are neither born modal, nor achieve modality, but have modality thrust upon them.

With this strategy in mind, consider Hempel's proposal (1965: 231): 'By a general law, we shall here understand a statement of universal conditional form which is capable of being confirmed or disconfirmed by suitable empirical findings.' The characterization has two parts: a specification of the syntactic form of a law and the stipulation that a law must be capable of being empirically disconfirmed. The latter clause rules out purely logical laws, while the syntactic specification requires a law to be expressible as a universal conditional.[18]

Hempel initially made this proposal in 1942, prior to the developments in modal logic that took place in the 1960s and 70s. The conditional he had in mind was the material conditional; laws were to have the form '$\forall x(Px \supset Qx)$'. If P and Q are extensional predicates, there is no modality contained in such a statement; it is equivalent to 'Nothing has property P but not property Q'.[19] But, while all genuine candidates for lawhood may be expressible in this form, so are many other statements, as Hempel himself became acutely aware. The statement, 'All the coins in my pocket are American currency', for example, can be disconfirmed, and a logician would render it as a universal conditional. None the less, although it looks like a law, it is not a candidate for lawhood.[20]

---

[17] Indeed, the term 'explanation' does not appear in the index of his book.

[18] The logical equivalence of $\sim\exists xPx \& \sim Qx$ and $\forall x(Px \supset Qx)$ compels us to state the requirement in terms of expressibility.

[19] A more restricted version is '$\forall x(Pxt \supset \exists yQyt')$' where t and t' are times, not necessarily distinct. For present purposes, however, the case given in the text is not misleading.

[20] The use of 'lawlike' to describe statements that are genuine candidates for lawhood is standard in the literature. This is unfortunate, since the problem with spurious candidates is precisely that they too look just like laws.

Why is this? One might perhaps rule it out on the grounds that it makes crucial reference to a specific entity—to wit, my pocket—and thus lacks the generality of a law. If this criterion is applied, however, then there can be no laws of cosmology, like Hubble's Law, since by definition there is only one cosmos. And, if we think that the Bode–Titius formula was just a useful summary of facts about certain planets and the asteroid belt, rather than a law, this is not, surely, just because of its implicit reference to our own solar system.[21]

To drive the point home, I quote van Fraassen on a well-known pair of examples:

(1) All solid sphere of enriched uranium (U^{235}) have a diameter of less than one mile.

(2) All solid sphere of gold (Au) have a diameter of less than one mile.

Of these he writes:

I guess that both are true. The first I'm willing to accept as putatively a matter of law, for the critical mass of uranium will prevent the existence of such a sphere. The second is an accidental or incidental fact—the Earth does not have that much gold, and perhaps no planet does, but the science I accept does not rule out such golden spheres. Let us leave the reasons for agreement to one side—the point is that, if (1) could be a law, if only a little law, and (2) definitely could not, it cannot be due to a difference in universality.   (1989: 27)

This problem is not specific to Hempel's characterization of laws. More fundamentally, it shows that our new strategy has simply exchanged one task, that of justifying the inference from universality to necessity, for another, that of selecting from the set of universal conditionals those that should properly be called lawlike. In short, our strategy has led us to the second problem, that of distinguishing laws from accidental generalizations.[22]

The standard answer to this problem is to say that laws license subjunctive conditionals, whereas accidental generalizations do not. Kepler's third law of planetary motion, put forward in *Harmonice Mundi*, provides a fictitious illustration of the criterion at work. For each planet (including the Earth)

---

[21] 'The mean distances of the planets from the Sun obey an exponential "law" discovered by Titius in 1776, and later confirmed and published by Bode. It has the form: $A = a + 2^n b$, where $a = 0.4$; $b = 0.3$; and n runs through $-\infty, 0, 1, 2, 4, 5, 6$.' This entry appears in Motz 1964: 181. The use of scare quotes is revealing.

[22] As Ayer pointed out (1956/1969), and my example shows, the term 'accidental generalization' is something of a misnomer. Though it is hardly a law of nature, it is no accident that, as I sit in front of a word processor in Columbia, South Carolina, all the coins in my pocket are American currency.

a remarkable relation held between its period of orbit (T) and its mean distance from the Sun (R). In each case, the ratio T^2/R^3 was the same. For metaphysical reasons (previously expounded in *Mysterium Cosmographicum*), Kepler believed that there could only be six solar planets, but we can imagine a later astronomer—call him Schmerschel—unencumbered by that belief, who surmised that, were there a seventh planet whose period of orbit was eighty-four Earth years, then the radius of its orbit would be roughly nineteen times as great as the Earth's. In doing so, Schmerschel would have endorsed two claims: the claim that Kepler's third law is indeed a law and the claim that a law licenses subjunctive conditionals of the form: 'If F *were* the case, then G *would be* the case'. In contrast, accidental generalizations do not license such conditionals. Were a Canadian looney to be placed in my pocket, it would not turn into a Susan B. Anthony dollar.

The criterion certainly accords with our (philosophically honed) intuitions. Yet, if we ask how it is that laws can license or support subjunctive conditionals, we may become uneasy. For, on our best theory of these conditionals, their truth values depend on how things are in worlds other than our own.[23] Our analysis, it turns out, has resolved the second problem at the price of returning to the first.

Recall that, in order to explicate the relationship of laws to physical necessity, we turned to the 'privileged class' strategy; we first denied that laws were inherently modal, and next defined physical possibility in terms of laws. The notion of lawhood, on this account, was prior to that of physical necessity. That strategy in turn required a criterion whereby genuine candidates for lawhood could be distinguished from other universal conditionals; the criterion selected was the capacity to support subjunctive conditionals. But, if laws have this capacity, then they must give us information about worlds other than our own—that is, they must have modal force. It appears that our criterion presupposes the very thing that we explicitly denied.

3.1.3. *Ceteris Paribus* Clauses

Variants of the phrase 'other things being equal' occur in many laws. Ohm's Law, for instance, tells us that, other things being equal, the ratio of the current flowing through a conductor to the potential difference between its ends is constant. We are now faced with a dilemma. If we omit the clause,

---

[23] See Stalnaker 1968 and Lewis 1973.

the statement is false. If other things are not equal—if, for example, the temperature of the conductor changes—then so does the ratio. If, on the other hand, we include the clause, the statement is empty. It says that unless something happens to upset the ratio it will stay the same.[24]

The radical response to this trio of problems is to say that, while no coherent account of lawhood can be given, this is not a philosophical tragedy, since laws themselves are otiose. This is the response given by Bas van Fraassen. The laws whose usefulness he challenges, however, are 'Laws of Nature'. In this essay I will confine myself to the laws of physics. Though less emphasized than it once was, the term 'law' is still part of the physicist's working vocabulary.

In sketching the problems encountered by the concept of laws, I have followed the philosophical literature which, by and large, assumes that there is a generic concept of lawhood, prone to generic problems. One symptom of that shows itself in the generic appellation 'Laws of Nature', which is found in the titles of philosophical papers; though these papers are collected in anthologies in philosophy of science, very little attention is paid to the sciences themselves.[25] In this essay I will confine myself to the laws of physics. Even in this genus, however, there are different species to be found.

In what follows, I distinguish three different categories of laws: Theoretical Laws, Conservation Laws, and Physical Laws. All three species play, or have played, a significant role in the practices of physics.

3.2. THE REPRESENTATIONAL ACCOUNT OF THEORIES

3.2.1. Empirical Generalizations and Theoretical Laws

On a widely held view of scientific theorizing, theory construction takes place in three stages. The accumulation of individual observations is followed, first, by the inductive generalizations that yield experimental laws, and then, by the articulation of theory.

---

[24] On falsity versus emptiness, see Cartwright 1983: 45–6. She writes: 'The literal translation is "other things being equal"; but it would be more apt to read "ceteris paribus" as "other things being right".' A discussion of provisos in general is given in Hempel 1988.

[25] In 1960, Stephen Toulmin deplored the choice, endemic in philosophical circles, of 'ravens are black' as an example of a scientific law ([1953] 1960, 10); nowadays the comparable choice is 'aspirin cures headaches'.

The distinction between experimental generalizations and theoretical laws was not a syntactic one. Even if it transpired that Hempel's syntactic characterization of a law was not quite apt, the required modifications of it would apply alike to theoretical and empirical laws. Nor is the distinction a semantic one, if that is taken to involve a distinction between observational and theoretical predicates.

Two paradigm cases will serve to emphasize this point. The first is the relation established by Kammerlingh Onnes between the transition temperature T at which superconductivity occurs in a conductor and the ambient magnetic field B. Onnes discovered that, in the absence of a magnetic field, superconductivity occured in a given metal only at or below a specific temperature T_C. If a magnetic field B was applied, however, then the transition temperature T fell; indeed, if B was above a threshold value B_0, superconductivity never occured at all. Onnes' experiments showed B and T to be related by the equation:

$$B = B_0 \left[1 - (T/T_C)\right]$$

This was a true (or approximately true) relation between two theoretical magnitudes, temperature and magnetic field. None the less, the equation was an experimental generalization.

In contrast, consider Newton's first law of motion, as he, or his translator Motte, states it: 'Every body continues in its state of rest or of uniform motion in a right line, unless it is compelled to change that state by forces acting upon it (1686)'. A logical empiricist would class all the predicates in the main clause as observable predicates, and even the theoretical term 'force' that appears in the proviso has a phenomenological correlate. Yet, for both conceptual and historical reasons it would be an empirical generalization which was later granted the status of a law. It is a paradigm of a theoretical law, not least because it is one law of three.

The difference in kind between the Onnes relation and Newton's first law may seem obvious enough, but other cases are less clear cut. In fact, the same statement may appear at one time as an experimental generalization and at another as a theoretical law.

3.2.2. The Representational Account of Physical Theory

Hilbert in 1900 saw the axiomatization of physics as one of the chief tasks awaiting twentieth-century mathematics, and the 'statement view of theories', as it became known, remained standard for philosophers of physics another sixty years.[26] Since then, it has come under severe attack. At the end of this section, I will add to these criticisms by showing that the statement view faces a fundamental problem, but, until then, I will be concerned to articulate and defend an alternative. The account I favour is a variant of the so-called *semantic approach* to theories offered by Ronald Giere. On this account, proposing a physical theory involves two things: (a) the specification of a class of models (the *theoretical definition*) and (b) the hypothesis that a certain part of the world, or of the world as we describe it, can be adequately represented by a model from this class (the *theoretical hypothesis*). The emphasis on models is due to van Fraassen (1980: chap. 3), the distinction between definition and hypothesis to Giere (1985: 78–80). Where Giere requires the subject of the model to be 'similar to the proposed model in specified respects and to specified degrees', I characterize the relation between model and world as one of *representation*. Like Nelson Goodman (1968: 3–10), I reject the view that a representation must be similar to its subject. In this essay, I explicate the crucial terms 'model' and 'representation' by giving examples, and postpone a fuller account to Essay 5. I call this account the *representational account* of theories, for these reasons: in part to emphasize what I take to be the crucial element of this approach, in part to differentiate this account from others which dwell under the semantic tent, but, more generally, from a growing conviction that the phrase 'the semantic approach' is at best obscure and at worst misleading, since one announced reason for adopting this approach is to move the philosophy of physics away from a preoccupation with questions of language.[27]

Pierre Duhem too veered away from explanations, and towards representation. In his *The Aim and Structure of Physical Theory* he showed how to construct his ideal theory. Here is the first paragraph of the introduction, and the first of four 'operations' in Part One, Chapter II:

A physical theory is not an explanation. It is a system of mathematical propositions, deduced from a small number of principles, which aim to represent as simply,

---

[26] See the first section of Essay 1.

[27] See van Fraassen 1980: chap. 3. Another desideratum is to distance the representational account from the set-theoretical articulations of theory to which the phrase 'the semantic approach' is also applied. See, for example, Sneed 1977 and Stegmüller 1976. This is the approach I refer to as 'the structuralist view of theories', in Essay I, where I compare it with the statement view; see n. 9 of that essay.

as completely, and as exactly as possible a set of experimental laws. (Duhem *PT*, 19)

...

1. Among the physical properties which we set ourselves to represent we select those we regard as simple properties. ... These mathematic symbols have no connection of an intrinsic nature with properties they represent; they bear to the latter only the relation of sign to thing signified. Through methods of measurement we can make each state of a physical property correspond to a value of the representative symbols, and vice versa. (Duhem *PT*, 20)

Note the occurrences of 'represent' and 'representative'. Echoes of those words sound in Part Two, Chapter III, where Duhem talks about an experiment in which 'we submit on a block of ice to a certain pressure', and 'the thermometer reads a certain number'. Two descriptions are presented. There are 'theoretical facts' and 'practical facts':

In such a theoretical fact there is nothing vague or indecisive. Everything is determined in a precise manner: the body studied is geometrically defined; its sides are true lines without thickness, its points true points without dimensions; the different lengths and angles determining its shape are exactly known; to each point of this body there is a corresponding temperature, and this temperature is for a number not to be confused with any other number.

Opposite this *theoretical* fact let us place the *practical* fact translated by it. Here we no longer see anything of the precision we have just ascertained. The body is no longer a geometrical solid: it is a concrete block. However sharp its edges, none is a geometrical intersection of two surfaces; instead, these edges are more or less rounded and dented spines. Its points are more or less worn down and blunt. The thermometer no longer gives us the temperatures at each point but a sort of mean temperature relative to a certain volume whose very extent cannot be exactly fixed. (Duhem *PT*, 132–3)

The theoretical fact is a representation of the practical, or physical, fact.

3.2.3. *Ceteris Paribus* and the Representational View

Armed with the representational account of theories and the role of laws within them, we may return to the problem of *ceteris paribus* clauses.[28] The problem took the form of a dilemma: either we omit the 'other things being

---

[28] I find that a broadly similar treatment has been given by Giere 1988, which is less than surprising.

equal' clause from the statement of a law, thus rendering the law liable to being false, or we include it, thus rendering the law empty.

Though he did not state it in exactly these terms, Duhem was much exercised by this problem.[29] His solution is to regard the laws of physics as neither true nor false, because, 'Every physical law is provisional because it is symbolic'. (That is the title of section 4 of Chapter V in Part II.) For an example (call it 'L_0'), consider the equation linking the density, temperature, and pressure of oxygen. If the sample of the gas is placed between the plates of a strongly charged condenser, the equation has to be modified by introducing a new variable (the intensity of the electric field between the plates) to produce a new law, L_1. If then an electromagnet, with its magnetic field perpendicular to the electric fields of the condenser, is switched on, L_1 will be falsified. Another law, L_2, is called for to take into account the magnetic field, but, even if such a law is forthcoming, there is another lurking in the bushes. In contrast, on the representational account of theories, the question of truth is finessed. On this account, a theory presents us not with a set of statements but with a model (more precisely, with a class of models). Unlike statements, models are neither true nor false. Instead, they offer more or less adequate representations of the world. By definition, the laws which appear within a theory are true in the models that the theory defines.

Take the case of Ohm's Law, which is normally stated without a *ceteris paribus* clause. As an experimental generalization, it is false. True, under controlled conditions, and within a considerable range of currents, we do indeed find a fairly strict proportionality between current and potential difference. Outside these limits, however, the law breaks down. If we raise the temperature of a conductor, the current–p.d. ratio will decrease; if the current is sufficiently large, the heating effect of the current itself will do the same. (As will currents so small that the nature of electric charge is discrete, and therefore cannot be a current.) Again, the application of a magnetic theory can produce changes in the current–p.d. ratio. Ohm's law, as a theoretical law appearing in our elementary theory of electric circuits (which models electric currents by analogy with the flow of an (ideal) homogeneous fluid through a pipe), is true by definition.[30]

[29] See chap. V. 3 in Part One of *PT*.

[30] This is not the whole story. Around 1825, when Georg Ohm started his investigations into 'galvanic current', he first thought in terms of an analogy with the theory of hydrodynamics; he then changed horses, and in his published work, *The Galvanic Circuit* (1827), he drew on

This theory does much useful work; to Ohm himself it suggested new concepts like that of the internal resistance of an electric cell. Moreover, as I pointed out, over a considerable range of currents and in a great variety of circumstances, the models it provides are adequate to our purposes.

Carl Hempel, however, contested the value of this approach. He wrote:

[T]he perplexities of the reliance on provisos cannot be avoided by adopting a structuralist, or non-statement, conception of theories. ... [T]he systems [presented on these accounts] make no assertions and have no truth values. But such systems are presented as having empirical models; for example, the solar system might be claimed to be a model of a structuralist formalization of Newtonian celestial mechanics. But a formulation of this claim, and its inferential application to particular astronomical occurrences, again assumes the fulfillment of pertinent provisos. (Hempel 1988: 31)

This argument was explicitly directed against the structuralist conception of theories proposed by J. D. Sneed (1977) and Wolfgaing Stegmüller (1976: esp chap. 7).[31] It is not obvious what force it has against the representational account offered here. Any account of theories, surely, must include the requirement that they be empirically adequate, though what counts as empirical adequacy will vary from theory to theory, and from application to application. The representational account is no exception. Empirical adequacy is a major, arguably the most important, component and as such is fully recognized in the requirement that the theoretical definition of a model be accompanied by a theoretical hypothesis. The model may be rejected. Less radically, we may settle for an approximate fit between model and world, or we may adjust the model to allow for perturbations, or, when deviations from theoretical behaviour are too great to be treated like that, we may create a new model that accommodates the original phenomena and the disturbing influence on equal

the mathematical treatment of the flow of caloric through thermal conductors published by Jean Baptiste Fourier in 1822. For more detail on Ohm's work, see Jungnickel and McCormmach 1986 vol. i. They tell us (Jungnickel and McCormmach 1986: i. 53): 'In 1830 Ohm remarked that if he had more success than others in explaining certain phenomena, it was because he was constantly guided by "theory".'

31 The Sneed–Stegmüller versions of the semantic view of theories were much more formal that the representational account I offer here; theoretical systems were exhibited as set-theoretical structures, and, as Hempel points out, inadequate attention was given to the fit between those mathematical structures and the world. Nor, as I have said, do I offer much on the model-world relation in this essay; I postpone that to Essay 5. Note that in the Sneed–Stegmüller accounts, the term 'model' has a different referent than it has in mine. As Hempel indicates, Sneed and Stegmüller both provide set-theoretical models of, for example, generic quantum systems or classical systems, rather than of aspects of the physical world.

terms. (Think, for example, of a theoretical representation of the electric pendulum.)

On the representational account, we should regard the qualification of a law by a proviso as a *façon de parler*. Strictly speaking, the limitation of scope implicit in the proviso does not apply to the law itself, but to the models whose behaviours the laws describe. A theoretical law is always true in the relevant model, but that model may not always adequately represent the parts of the world that interest us. This is the sense in which we should understand Toulmin's remark about laws: 'They are the sorts of statement', he writes ([1953] 1960: 86), 'about which it is appropriate to ask, not "Is this true?" but rather, "To what systems can this be applied?" '.

The account of theoretical laws that I have given finds support in Feynman's presentation of Kirchoff's laws of electric circuits (Feynman, Leighton, and Sands 1963–5: ii. 22.8). These, too, are theoretical laws from which *ceteris paribus* are customarily omitted. Feynman sets them out in algebraic form, as follows:

For any closed loop $\Sigma V_n = 0$ For any node, $\Sigma I_n = 0$
of a circuit,   *around loop*   *into node*

(V_n refers to the potential drop across any one of the elements in the elements in a closed loop; I_n refers to the current entering a node via a specific conductor.)

Although no *ceteris paribus* is stated, these laws run no risk of being false. This section of Feynman's lecture is headed, 'Networks of ideal elements', and he prefaces his statements of the laws with the remark: 'By making the many approximations we have described in [the previous section] and summarizing the essential features of the real circuit elements in terms of idealizations, it becomes possible to analyse an electric circuit in a relatively straightforward way.' In other words, the laws as stated are laws of the behaviour of a class of models, which for certain purposes provides adequate representations of actual electric circuits. Here the account I have given of a scientific theory, and of the place of laws within it, finds striking confirmation within scientific practice.

Feynman's use of the phrase 'ideal elements' prompts the following reflection. Elements of actual circuits are manufactured rather than found in nature. In view of that, perhaps Kirchoff's laws should be thought of not as expressing empirical regularities but as providing a yardstick by which the performance of those artefacts can be measured. I will not pursue this issue here.[32]

3.2.4. The Modal Force of Laws

I turn now to problems involving modality. First, what kind of necessity, if any, does a law express? Secondly, how does a law license subjunctive conditionals?

If we view a law as an experimental generalization, then it is very hard to attribute modal forces to it. To the justification that I gave in section 3.1.1, we can add the problem of meaning: if we add a modal operator to an empirical generalization to produce a modal law, what exactly does that operator signify? If all Ps—past, present, and future—are (tenselessly) Qs, what is added when we say that it is necessarily so?

If, however, we consider theoretical laws and adopt the representational account of theories, these problems disappear. Recall that on this account, a theory specifies a class of models, one of which is taken to represent our world. In doing so, the theory also picks out a set of possible worlds, the worlds that are representable by the models in that class. For Newton, it was important that his theory represented a world in which six planets orbited the Sun, but, other possible worlds, in which there are seven, eight, nine, or a multitude of planets, are also representable by the theory. The theoretical laws—that is, the statements that either define the class of models or follow from those definitions, are true in all models within the class. It follows that they apply to all theoretically possible worlds, and so express, theoretically, necessity. Hence their modal force, and their capacity to license subjunctive conditionals.

This may be set out more precisely. If we read Kepler's laws as they appear in Book I of the *Principia*, that is, as conditionals specifying the kinematic behaviour of non-interacting bodies moving inverse square centripetal fields of force—in brief, if we view them as theoretical laws—then they are modal statements, in that they are true in all models of Newtonian theory. This is the modality to which we appeal when we say that the theory licenses subjunctive conditions about solar systems. Nor is this appeal inappropriate, provided we recognize that it requires a stronger version of the theoretical hypothesis

---

[32] Co-sponsor to my thought was Cartwright et al. 1994, which I read while working on this section of the essay.

than the one stated hitherto. To use the language of the possible-worlds theorist, the hypothesis must be that, not only is our own world adequately represented by the theory, but so are other possible worlds; furthermore, that in the space of possible worlds, these particular worlds lie very close to our own.[33]

3.3. LAWS OF NATURE

3.3.1. Lewis on Laws of Nature

The kind of necessity we have arrived at may be not as strong as some were hoping for. It is a theoretical necessity, or rather a family of theoretical necessities, each of them physical necessity as represented within a particular theory, none of them physical necessity *tout court*. Given the limitations of theoretical scope acknowledged in the discussion of *ceteris paribus* clauses, it is clear that the laws of physics, as analysed here, are a long way from the laws of nature.

This raises the question: *Are* there any laws of nature? We know there are laws of physics, since we find them in physics textbooks, but (*pace* Galileo) the subtext of the book of nature does not lie open to our gaze. If laws of nature exist, they will (presumably) transcend individual theories. But, in that case, how does science provide them, since theories are all that science can give?

To link theoretical laws with laws of nature, while recognizing that in the present state of our knowledge they do not coincide, would be to look forward to some all-embracing theory, the asymptotic limit of scientific theorizing. The laws of such a theory could then be regarded as laws of nature.

A proposal along these lines has been made by David Lewis.[34] It offers, *inter alia*, an answer to the question raised in section 3.1.2 about the criteria for lawhood. Lewis writes (1973: 73): 'We can restate Ramsey's 1929 theory of follows: a contingent generalization is a *law of nature* if and only if it appears as a theorem (or axiom) in each of the true deductive systems that achieve a best combination of simplicity and strength.' Lewis does not

---

[33] I employ here the vocabulary of Lewis's account of counterfactuals; see chap. 1 of Lewis 1973.
[34] Ibid. 72–7. Note, however, Lewis's disclaimer: 'I doubt that laws of nature have as much of a special status as has been thought' (ibid. 73).

say that a theory of this kind would appear as the limit of a succession of successful theories that included those we presently accept. He does claim, however (1973: 74), that 'Our scientific theorizing is an attempt to approximate, as best we can, the true deductive system with the best combination of simplicity and strength'. But, in that case, in making these attempts, physicists have been—and are doomed to be—content with a very low-level approximation, if, indeed, that word is appropriate at all. For, as we saw in section 3.2.3, if 'true' means 'literally true', then the generalization of our best theories neither is, nor can be, set out as a deductive system, for the reasons explored there. The apparently innocuous phrase, 'true deductive system', contains crucial but ultimately indefensible assumptions about the applicability of the geometrical method to physical theories. A complete and true system of the kind proposed could act as a regulative ideal for physics only in the way that the Fountain of Youth acted as a regulative ideal of Ponce de Leon.[35]

3.3.2. Einstein and Noether on Laws of Nature

Lewis's account of laws of nature, though articulated with great subtlety, suffers from an uncritical reliance on the statement view of theories. His definition, in which he specifies the conditions under which 'a contingent generalization is a law of nature', suggests a view similar to Duhem's. Recall that in 1914 the latter defined *theory* as 'a system of mathematical propositions, deduced from a small number of principles, which aims to represent as simple, and as exactly as possible a set of experimental laws'. (*PT*, 19). Duhem wrote this, not as a dispassionate observer of the physics of his day, but as a protagonist in the debate between energeticists and atomists over the proper nature of physical theory in general, and thermodynamics in particular.[36] In his characterization of the atomic hypothesis as 'metaphysics' (*PT*, 12–13), as in his assertion that 'the validity of a theory is measured by the number of experimental laws it represents, and by the degree of precision with which it represents them' (*PT*, 288), he declared himself a confessed energeticist. For the energeticist, the kinetic theory of gases might furnish a useful way to summarize phenomenological laws, but had neither

---

[35] A scrutiny of physics journals suggests that, to a great degree, physicists are content to mix and match theories to phenomena, rather than to seek a unified theoretical approach to them. See Essay 4.

[36] On the energeticist debate, see Jungnickel and McCormmach 1986: ii. 21–7.

explanatory value (for explanatory power was not a value), nor any claim to provide an adequate characterization of the constitution of matter in the gaseous state.

As it turned out, the energeticists were on the losing side. On the specific issue of atomism, the experiments and writings of Jean Baptiste Perrin (1913) convinced the majority of the scientific community of the reality of atoms. At the same time, experiments, like those of Hans Geiger and Ernest Marsden in 1909 that scattered alpha-particles by thin films provided new and remarkable phenomena rather than experimental laws.[37] These phenomena were not to be accommodated within abstractive theory, but in the theory of the constitution of the atom that Bohr supplied in 1913. No argument was needed to show that this theory specified a model of the model of the atom, and, after a short time, it was accepted as a *bona fide* representation of the atoms occurring in nature.[38]

In the same year, Moseley's Law, one of the few twentieth-century laws of physics, was announced. It is an experimental law that relates two theoretical quantities, the frequency f of a particular line in the X-ray spectrum of an element, and the element's atomic number, Z. We have:

$$f = [a(Z - b)]^2$$

(The coefficients a and b depend on the particular spectral line being examined.) The law shows how far scientific practice had diverged from Duhem's account of physical theory. No attempt was ever made to coordinate this law with others in an abstractive theory. Instead, physicists looked to it for information about atomic structure, in particular about the charges on the atomic nuclei of different elements.[39]

Throughout that century, the thrust of experimental physics was the investigations of matter and the forces that bind it together. And, such regularities as have been found have not been organized under laws. More 'effects' than 'laws' have been given the names of physicists. But, while much of physics has been 'inward bound' in Pais' nice phrase, the century also saw a revolution in our picture of time and space. Again, the discovery of new laws was not part of this revolution; nevertheless, the rules that Einstein appealed to in articulating the special and the general theories of relativity are

[37] Wehr and Richards 1960.   [38] See Hughes 1990b.
[39] See Wehr and Richards 1960: 163–640; Pais 1983: 228–30.

centred on the notion of law. 'The universal principle of the special theory of relativity', he writes ([1915] 1949: 57), 'is contained in the postulate: The laws of physics are invariant with respect to the Lorentz transformations'. He later generalized the rule to the principle underlying the general theory (1949: 69), but I will not consider that extension here.

The laws examined by Einstein in his 1905 paper, and which yielded the special theory, were those expressed by Maxwell's equations. In that analysis, Einstein worked from the laws of electromagnetism to the Lorentz transformations. It is clear, however, that his principle, once accepted, becomes a criterion of lawhood. The laws of our fundamental theories are to be invariant under specific coordinate transformations; to put it another way, they are to conform to certain symmetry requirements.

Observe (1) that this criterion transcends individual theories. It applies alike to classical electromagnetic theory, relativistic classical mechanics, and quantum field theory. (Ordinary quantum mechanics does not fulfil it: that has been a gnawing problem within physics since the 1920s.) Note also (2) Emmy Noether's demonstration that those conservation principles which we regard especially significant, and which reappear in a variety of different theories, are all tied to specific symmetries (Kaku 1993: 23–30). For instance, in relativistic classical mechanics, three principles of conservation hold: the principle of energy conservation (provided time is independent), of linear momentum (provided translation is independent), and of angular momentum (provided rotation is independent). Similar results hold in classical electromagnetic theory and quantum field theory. In addition, note (3) that from the 1930s onwards symmetry considerations have played an ever-increasing role in particle physics.

The first two of these considerations, and perhaps the third, suggest that the concept of symmetry might offer the kind of generality required of laws of nature. Before jumping to conclusions, however, we should pause to reflect that, though they may underlie many theories, these symmetries, like the laws discussed in this essay, are themselves theoretical. They are the symmetries postulated by our deepest theories, notably our theory of space-time. They remain, therefore, symmetries of our representation of nature; we cannot simply assume them to be symmetries of nature itself. I postpone discussion of them to a future occasion, and here content myself with just one remark. To talk in terms of symmetries rather than laws would, at the least, obviate one kind of misconception: whereas laws are a species of statement, symmetry is a property of models.

3.4. NEWTON'S *PRINCIPIA*
AND THE REPRESENTATIONAL ACCOUNT
OF THEORIES

A virtue of the representational account of physical theories outlined in section 3.2 is that we can claim not only that it corresponds to the theoretical practices of both the nineteenth and twentieth centuries but also that all post-Galilean physics is illuminated when viewed in this light. Since the twentieth century and its successor have seen few new laws of physics, for present purposes the second claim is the more important of the two. It is also the most contentious, for obvious reasons: for 200 years, from Newton to Duhem, mathematical physicists advocated a purely Euclidean presentation of physical theories.

I will grasp this nettle by taking up the most celebrated of all works of mathematical physics, Newton's *Principia.* My aim is twofold: to show that the representational account gives an accurate picture of Newton's theoretical practice, and to find a place in that account for theoretical laws. To keep the discussion within bounds, I will examine those passages that deal with Kepler's laws of planetary motion. They occur in two segments of the work: first, the opening sections of Book I, containing the Definitions and Axioms, together with section II and (half of) section III (*P*, 1–62), and secondly the early parts of Book III, up to Proposition XIII (*P*, 397–420).[40]

This particular nettle turns out to be remarkably soothing to the hand. The two segments neatly illustrate Giere's bipartite characterization of a theory. In Book I, Newton examines the models specified by his theoretical definitions; in Book III, he deploys his theoretical hypothesis.

As Newton himself says, he lays down principles in the first part of the *Principia* which are 'not philosophical but mathematical' (*P*, 397).[41] He begins by setting out eight *Definitions*, each accompanied by a commentary. The first two define *quantity of matter* (or *mass*), and *quantity of motion*, recognizable as our quantity of momentum (*P*, 1). The third defines *vis insita*, or innate form of matter, recognizable as our concept of inertia

---

[40] Throughout the remainder of this essay, I will abbreviate Newton [1726] 1934 to *P*.
[41] 'Philosophy' here is the *philosophia naturalis* of the work's title.

(*P*, 2): 'The *vis insita* is a power of resisting, by which every body, as much as in it lies, continues in its present state, whether it be of rest, or of moving uniformly forwards in a right line.' The fourth speaks for itself: 'The *impressed force* is an action exerted on a body, in order to change its state, either of rest, or of uniform motion in a right line' (ibid.).

These definitions are perfectly general. The remaining four are all concerned with centripetal forces. *Centripetal force* is defined in Definition V, and definitions of *the absolute, accelerative*, and *motive quantities of a centripetal force* are given in Definitions VI, VII, and VIII. Each definition is followed by some illustrative remarks (*P*, 2–4).

Definition V. A centripetal force is that by which bodies are drawn or impelled, or any way tend, towards a point as to a centre.

Of this sort is gravity, by which bodies tend to the centre of the earth; magnetism, by which iron tends to the loadstone; and that force, whatever it is, by which planets are continually drawn aside from the rectilinear motions, which otherwise they would pursue, and made to revolve in curvilinear orbits.[42]

Definition VI. The *absolute quantity* of a centripetal force is the measure of the same, proportional to the efficacy of the cause that propagates it from the centre through the spaces round about.

Thus the magnetic force is greater in one loadstone and less in another, according to their sizes and strength of intensity.

Definition VII. The *accelerative quantity* of a centripetal force is the measure of the same, proportional to the velocity which it generates in a given time.

Thus the force of the same loadstone is greater at a less distance, and less at a greater: also the force of gravity is greater in valleys, less on tops of exceeding high mountains; and yet less (as soon shall be shown), at greater distances from the body of the earth; but at equal distances, it is the same everywhere; because (taking away, or allowing for, the resistance of the air), it equally accelerates all falling bodies, whether heavy or light, great or small.

Definition VIII. The *motive quantity* of a centripetal force is the measure of the same, proportional to the motion which it generates in a given time.

Thus the weight is greater in a greater body, less in a less body; and, in the same body, it is greater near to the earth, and less at remoter distances. This sort of quantity is the centripency, or propension of the whole body towards the centre, or, as I may

---

[42] Notice that, at the very beginning of the *Principia*, the same laws are to be applied to terrestrial and celestial bodies alike.

say, its weight; and it is always known by the quantity of an equal and contrary force just sufficient to hinder the descent of the body.[43]

After the *Definitions* and *Axioms* (with four Corollaries attached[44]) comes *Book I, The Motion of Bodies*, in which a few simple systems are examined. A Newtonian system is a model of the theory; it consists of a configuration of masses and forces whose behaviour conforms to the three laws of motion. Those examined in Propositions I–XIII of Book I all consist of one or more bodies moving within a centripetal field of force.[45] In each Proposition, the kinematic behaviour of the system is related to the field of force via the dynamical principles set out in the three Laws of Motions. Sometimes Newton argues from the kinematic behaviour to the nature of the force field responsible, and sometimes in the opposite direction, from the specification of the field of force to the trajectories of the masses. The main emphasis is on arguments of the first kind, from behaviour in accordance with Kepler's laws to the existence of a field of force whose strength varies inversely with the square of the distance from its centre (an 'inverse square field' we may call it). In these Propositions his strategy is dictated by a desire to show that the astronomical phenomena compel him to his law of gravitation, and he cites these phenomena in arguing for this law in Book III. The argument is completed in Proposition VII, Theorem VII of that Book (*P*, 414).

Of more relevance to our topic are the Propositions in which Newton shows that certain kinds of force fields give rise to behaviour conforming to Kepler's laws. Thus, in Proposition I, Theorem I (*P*, 40), Newton shows how Kepler's second law (the equal area law) is obeyed by a body moving in a centripetal field of force. In Corollary VI to Proposition IV (*P*, 465), he shows that, when a number of bodies move in circular orbits within a centripetal field, Kepler's third law is obeyed: the ratio T^2/R^3 has the same value for each body. In Proposition XV (*P*, 62), he shows that the same law applies in the case of elliptical motions in a centripetal field whose centre is a focus of the ellipses. The key result, that the trajectory of any body moving

[43] A page of commentary separates Definition V from Definition VI. I have quoted in full the illustrative remarks after Definitions VI, VII, and VIII. After these remarks, Newton gives a three-paragraph commentary on those definitions, before embarking on the *Scholium* (*P*, 6).

[44] Corollary III is the corollary which derives the conservation of momentum principle for a system consisting of interacting bodies obeying the Laws of Motion.

[45] The term 'field' is anachronistic but convenient. Note, in particular, that Newton has defined a system where the 'accelerating quantity' of the force at a given point is the same whatever the mass of the object placed there (see Definition VII), as happens, for instance, within a gravitational field.

in an inverse square field will be a conic section with a focus at the field's centre, is tucked away in Corollary of Proposition XIII (*P*, 61). This result yields Kepler's first law as a special case.

I talk of 'Kepler's Laws', but, in the form in which Newton derives them, they carry reference neither to the planets nor to the Sun. The bodies whose trajectories are described are point masses, and the centres of the fields of force are geometrical points. Furthermore, when more than one body is involved, as in Propositions IV and XV, the bodies do not interact. The laws they obey are conditionals: given an inverse square centripetal force field, they tell us, the trajectories of these masses will be conic sections with the centre of the field as focus; the equal area law will hold; and the ratio T^2/R^3 will be the same for each body. These are theoretical laws, laws governing the models that Newton's theory defines.

I turn now to Book III. In the first Proposition of this Book, Newton claims (*P*, 46): 'That the forces by which the circumjovial planets are continually drawn off from rectilinear motions, and retained in their proper orbits, tend to Jupiter's centre; and are inversely as the squares of the distances of the places of those planets from that centre.' He adds, 'The same thing we are to understand of the planets which encompass Saturn' (ibid.). Propositions II and III make similar claims about the planets of the solar system, and the Moon in its orbit round the Earth.

These claims are all supported by appeals to Phenomena, and to salient Propositions from Book I. Newton has set out the Phenomena in the preceding pages (*P*, 401–5); he has presented astronomical data to show that the planetary systems in question obey the equal area law and the 2/3 power law (or, in the case of the Moon, that its apogee is very slow). In the Propositions he cites from Book I he has demonstrated the converses of Kepler's laws by arguments of the first kind: he has argued from the kinematic behaviour of the model to the nature of the field of force responsible. He has shown (1) that if the equal area law is obeyed, then the body in question is in a centripetal field, and (2) that if the 2/3 power law is obeyed (or if the apogee is very slow), then the field is an inverse square field.

As I have mentioned, for Newton these are mathematical principles 'such, namely, as we may build our reasonings upon in philosophical [i.e. scientific] enquiries' (*P*, 397). But, if these reasonings are to be cogent, Newton must assume that the mathematical principles are indeed applicable to the phenomena. This assumption, everywhere present in Book III of the *Principia*, is nowhere stated. As they stand, Newton's arguments are all enthymemes. In each case the suppressed premiss is an instance of what, following Giere, I have called the *theoretical hypothesis*, the postulate that a particular part of the world can be adequately represented by one of the models that the theory defines; stated generally, that there are models of the theory that are also models of the phenomena.

Thus, the representational account of theories makes explicit what is implicit in Newton's practice, and without which that practice would be incoherent. In sum, the evidence from the *Principia* confirms the adequacy of this account of theories in representing the phenomena that fall within its own ambit—that is, physical theories; furthermore, within it a place has also been found for theoretical laws: a theoretical law is a statement true in each member of the class of models that the relevant theory defines.

On my formulation, the form of a theoretical hypothesis is not that a physical system *is* a system of the type defined by the theory, nor even that it is *similar* to such a system (in specified respects, or to a specified degree).[46] Rather, it states that one can be *represented* by the other. A fundamental reason for insisting on this is, in this context, anachronistic. The physics of the twentieth century has shown that even a hugely successful theory like Newton's may be challenged by one with a very different ontology. *Pace* Descartes,[47] we can never claim that our present theory gives us moral certainty about the actual workings of the world; adequate representations are the most we can hope for. Yet, the case of Newtonian mechanics also demonstrates that a theory can be false but yet adequate for many purposes—witness the fact that physicists continue to use it to good effect.

This meta-theoretic reason aside, there is evidence enough in the *Principia* that the relation between the models described in Book I and the physical systems dealt with in Book III both is, and has to be, one of representation. An example will justify the weaker claim. In Proposition II of Book III (*P*, 406), Newton treats the planets of the solar system. He argues that the force field they move in varies as the inverse square of their distance from the Sun, by citing the result obtained in Book I for a system of bodies in circular orbits (*P*, 45). Yet, he knows that the orbits of the planets are not circular—indeed, the Phenomena he adduces in the same argument are facts about the 'mean

---

[46] The latter is Giere's preferred formulation.
[47] See §§205 and 206 in Part Four of Descartes' *Principles of Philosophy*.

distances' of planets from the Sun (*P*, 404). Thus, the suppressed premiss in his argument is that the theoretical model provides, not a true account, but an adequate representation of the phenomena.[48]

Now for the stronger claim. The 'System of the World' that Newton presents to us in Book III of the *Principia* is a system of massive bodies in continuous interaction with one another. These interactions are the resultants of pairwise mutual attractions governed by the Law of Gravitation. No other forces are involved. As a consequence of these forces the motion of each body is continuously changing; moreover, as the geometrical configuration of the system changes, so too do the forces the bodies experience and the accelerations they undergo. This is as true of the Sun as it is of any other body in the system.

As we have seen, Newton's arguments for his Law of Gravitation, and hence for the system as a whole, depend crucially on the mathematical principles derived in Book I. But, someone who adopted a literal, as opposed to a representational, construal of the relation between theory and world would regard these principles as irrelevant to the world we inhabit. For the literalist, each of these theorems would give us information about the system we have seen before, a system of non-interacting bodies moving in a centripetal force-field. No mass is associated with the cause of this force-field; that is one reason why Newton never needs to invoke his third Law of Motion in deriving these principles. The other is the absence of any interaction between the bodies. The field in each case is centred on a geometric point, which may be at rest or in uniform rectilinear motion. Not one of these conditions is satisfied by Newton's 'System of the World' in Book III. For the literalist, therefore, Newton's conclusions in that Book would invalidate the assumptions of the very argument on which they were based.

Similar consideration disclose a deep problem for the statement view of theories. Recall that on this view a theory presents us with a sequence of statements laid out in the geometrical manner. Now there are (at least) two ways in which geometry can serve as a paradigm for a physical theory. The force is the most obvious: '*More geometrico*' is now synonymous with 'using the axiomatic manner'. The second way, developed by Euclid and Apollonius, involves the methodology of traditional geometry, in which theoretical demonstrations involve diagrammatic constructions. Galileo, in

---

[48] Not until Prop. XIII (*P*, 420) does Newton acknowledge that the orbits of the planets are more accurately described as elliptical.

the first half of the seventeenth century, and Newton, in the second, were masters of this technique, but few, if any, have followed them.

Euclid, of course, combined both aspects of the Euclidean mode. One of the great strengths of the axiomatic method in mathematics is its cumulative nature. In Euclidean geometry we derive, first, comparatively simple configurations. We show, for example, that the internal angles of a triangle add upto two right angles. Then, by constructing a polygon as a set of triangles with a common vertex, we use the original result to prove a theorem about the internal angles of a polygon. The theorem proved for the simple configuration still holds when that figure is embedded within a more complex one. The facts about any given triangle within the polygon are unaffected by the presence of its neighbours.

In comparison, this cumulative method can rarely be applied within physical theories. A theorem proved for a simple configuration of forces and masses will, in general, not hold within a more complex one. Consider the theorems proved by Newton in the early sections of the *Principia*. As we saw, they were all proved for systems of non-interacting masses in a central field of force. It is true that an extra non-acting mass or two would not invalidate these theorems, but, if to the original configuration we add forces of mutual attraction between the masses, these theorems will no longer hold. The result is that a strict adherence to the geometric method would mean that, with every increase in complexity, the relevant results would have to be proved *ab initio*. We could not allow for these additional forces in the derivation simply by adding a few new premises to our deductions; the existing premises would have to be replaced. Indeed, only moderately complex systems would lie beyond the reach of theory; in celestial mechanics, for instance, the n-body problem is not solvable by analytical means for $n > 2$.[49]

To deal with the problems posed by mutual interactions between the bodies, Newton uses two complementary strategies, which we may call *strategies of approximation* and *strategies of perturbation*. In Book I, he shows how a system of bodies that themselves act as centres of inverse square fields of force can be approximated by his earlier centripetal model, provided that (1) the (absolute) forces associated with fields are proportional to the masses of the bodies,[50] and (2) most of the mass of the system is concentrated

---

[49] See Essays 4 and 6.
[50] A few pages later, in Prop. LXIX (*P*, 191), Newton invokes the third law of motion to provide a very elegant proof that 1 holds for such systems.

in a single body (Prop. LXV, Case I). In Book III (Prop. XIII), however, he acknowledges that departures from the centripetal model are sufficiently large to be measurable by the observational astronomy of his day. From the attractive force of Jupiter on Saturn, which Newton estimates to be about 1/112 that of the Sun, '[there] arises a perturbation of the orbit of Saturn in every conjunction of this planet with Jupiter, so sensible, that astronomers are puzzled with it' (*P*, 421).

Newton deals with this, not by working out Saturn's orbit from first principles, an impossibility that a strict adherence to the axiomatic would require, but by tweaking the centripetal model to get a better fit. The very notion of a perturbation, we may think, suggests a deviation from an ideal pattern, to be dealt with by an adjustment to the model in which the pattern is represented.[51]

The representational account of theories does not merely offer us pictures of these strategies that are closer to Newton's practice than any available on the axiomatic account. It accommodates with ease what, on the latter account, are insoluble problems.

3.5. HENRI POINCARÉ AND NEWTON'S LAWS OF MOTION

The most famous, or infamous, of all *ceteris paribus* clauses appears in Newton's first Law of Motion:

Law I. Every body continues in its state of rest or uniform motion in a right line, unless it is compelled to change that state by forces acting upon it. (*P*, 2)

Our suspicion that the final clause may make the law empty is reinforced when we recall Newton's earlier definition of a force:

Definition IV. An impressed force is an action exerted upon a body, in order to change its state, whether of rest or of uniform motion in a right line. (*P*, 2)

Before rushing to judgement, however, we should note two things. First, the proviso in question is not a run-of-the-mill *ceteris paribus* clause. A typical *ceteris paribus* clause leaves open the nature of the influences that may render the law false. In contrast, the great strength of Newton's theory is that he

---

[51] See the section on 'Deduction in the Bohm–Pines Quartet' in Essay 2.

brings all these influences under one heading. There are no 'other things' to be equal; deviations from uniform rectilinear motion are all to be attributed to forces. Secondly, within Newton's theory, reference to forces is not confined to Law I and Definition IV. Further information about them is supplied by his second and third laws of motion. A resolution of the proviso issue must therefore involve an appraisal, not just of the first law, but of the theory as a whole.

During the last two decades of the nineteenth century and the first decade of the twentieth, a number of physicists, notably Ernst Mach, Heinrich Hertz, and Henri Poincaré, undertook such appraisals. Of particular relevance to our concerns is the one provided by Poincaré; as will appear, there are striking affinities, but also marked contrasts between the analysis he provides and the account of Newtonian theory I outlined earlier. The problem that exercises Poincaré is that of the logical status of Newton's three laws. The first law, for instance, is neither experimentally falsifiable, nor is it an *a priori* truth. As Poincaré points out ([1902] 1952: 97), the presence of the proviso in the law guarantees that 'it can neither be confirmed nor contradicted by experience'. But, although the law is not a straightforward empirical truth, neither can it be justified on a priori grounds. A priori, the principle that the velocity of a body does not change without reason has no more and no less warrant than a similar claim about its position, or about the curvature of its path (ibid. 91). The logical status of the principle is thus unclear.

Prima facie, while Newton's first law describes a body's trajectory in the absence of a force, the second (the Law of Acceleration, as Poincaré calls it) tells us what happens when a force is applied: the body will undergo an acceleration proportional to the force and inversely proportional to the body's mass. The problem with this law, however, is that no independent measure of force is given; the only measure we have is provided by the relation $F = ma$.[52] Accordingly, the Law of Acceleration 'ceases to be regarded as experimental, it is now only a definition' (ibid. 100). But, 'even as a definition it is insufficient, for we do not know what mass is' (ibid.). The definition of mass Newton supplies, that mass is the product of the volume and the density, is circular. For (and here Poincaré quotes Thomson and

---

[52] In the *Principia*, Newton never uses algebraic notation. I do so here when it does not distort his argument. He uses geometry throughout, and the geometrical tradition which Galileo endorsed explicitly only used pure ratios; thus, in that tradition Newton's First Law would be rendered as 'F_1/F_2 is the product of m_1/m_2 and a_1/a_2'.

Tait), 'It would be better to say that density is the quotient of the mass by the volume' (ibid. 97).[53]

Poincaré resolves the problem by appealing to Newton's third law (the Principle of Action and Reaction). This law tells us that if, in an interaction between two bodies A and B the forces on them are F_A and F_B respectively, then $F_A = -F_B$. Whence, by the Law of Acceleration, $m_A a_A = -m_B a_B$, and so the ratio $-a_A/a_B$ of the bodies' accelerations gives the inverse ratio of their masses. But, then, like the Law of Acceleration, 'this principle of action and reaction can no longer be regarded as an experimental law but only as a definition' (ibid. 100). More precisely, 'Thus is the ratio of two masses defined, and it is for experiment to verify that the ratio is constant' (ibid. 101).

We see that, on Poincaré's analysis, the logical status of Newton's first law is moot, while the second and third are essentially definitions. Poincaré refers to them collectively as 'conventions'. We may compare his verdicts with those made from the standpoint of the representational account of theories.

The question that Poincaré starts from: whether Newton's first law is an empirical or an a priori truth, presupposes that a law is a *judgement*, in the Kantian sense. Poincaré's question is: What kind of judgement is it? In the early twentieth century, conventions were neo-Kantian stand-ins for Kantian synthetic a priori truths (as Michael Friedman has pointed out (1983: 18–22)). For Kant, the latter were judgements whose truth was guaranteed by the nature of our cognitive faculties. Poincaré's conventions, on the other hand, are true given the way we have decided to organize our account of the world. To take his most famous example, we may choose to describe physical space and time in terms of Euclidean geometry, at the risk of making our physics complicated, or to adopt an alternative geometry in order to make our physics simple. Our choice of a physical geometry is thus conventional, and the axioms of the geometry we use are conventions. The same applies to our choice of a theory of mechanics.

On Poincaré's view, then, Newton's laws are true given the way we describe the world; on the representational view, they hold, like all theoretical laws, in the models by which we represent the world. The accounts share a common feature: both contain an element of choice. On Poincaré's account we choose a particular way of describing the world; on the representational

---

[53] Here Florian Cajori, in a footnote to the *Principia*, quotes H. Crew: '[I]n the time of Newton density and specific gravity were employed as synonymous, and the density of water was taken arbitrarily to be unity.' This is entirely consistent with the geometrical tradition. See the previous note.

view we choose a class of models by which to represent it. The accounts are nevertheless significantly different. Poincaré take theoretical laws to be literal truths about the world, *modulo* a certain mode of description. In contrast, on the representational account, laws are true in the models that the theory defines. Indeed some of them constitute that definition.

Newton's laws of motion are a case in point. What Poincaré saw as a problem, that the second and third laws act as definitions, is, on the representational account, exactly what we would expect. Even the fact that, on its own, the second law acts merely like a definition, and the third is needed to turn appearance into reality, is not a significant defect. On the representational account, a theory is judged as a whole. When we assess, whether in terms of logical cogency or empirical adequacy, we look at the models of the theory rather than at individual statements describing them. This is precisely the kind of issue that distinguishes the representational account of theories from the statement view.

Of course, as successive clauses of the theoretical definition are introduced, we are free to stop at any stage to inspect the models defined by the clauses already in place and see what they might represent. When we examine Book I of the *Principia* with this in mind, a curious fact emerges. The theorems and corollaries proved in sections II and III, and put to such good use in Book III, require for their proof only the apparently empty first law, and the allegedly defective second law. How, we may ask, can this be, given Poincaré's demonstration that the second law is impotent without the third?

The answer suggested by the representational view is that in the early parts of Book I Newton has no need to provide an operational definition of force, since his enterprise is simply to define a class of models. The term can remain a purely theoretical one, to be cashed out in Book III, when the theory as a whole is related to the world. His cunning in Book I shows itself in the systems he selects for investigation. The centripetal fields he describes are such that the 'accelerative quantity' of the force at any point is the same, whatever mass it is acting on—which is to say that the fields behave like gravitational fields (see Definition VII). Hence, the masses of the moving bodies are irrelevant.

The moral of the story is that it is better to look at the models that a theory provides to than analyse the logical status of its individual laws.

To end this section where I began: What of the first law of motion? Is it as empty as it seems? From the logical point of view: yes. From the historical: no, for three reasons. First of all, it represents a major deviance from the accounts of natural motion given by Aristotle and Galileo.[54] Secondly, acceptance of the first law is a precondition for the acceptance of the second. Thirdly, the importance of certain novel phenomena in the lives of Newton's contemporaries may have made them peculiarly receptive to this particular law. These phenomena are representable by a model defined by the first law alone, as a brief excursion, or magical mystery tour, through the *Lebenswelt* of seventeenth-century England will disclose.

Hard is the task of one who would untangle the warp and woof of history, and perilous the enterprise of isolating the causal factors affecting the reception of such a work as Newton's *Principia*. Nevertheless, I will fly a kite for Clio, and pick out two: the invention of the game of billiards in the sixteenth century,[55] and the hard winters that prevailed in northern Europe in the latter part of the seventeenth.[56] For (to echo a point made earlier) the owner of the monopoly on the manufacture and sale of billiard tables will soon have learned that the price he could charge for his product varied directly with the time a ball would continue in rectilinear motion upon its surface. And, in 1666, *Anno Mirabilis*, when the Thames froze over, many a citizen will have discovered—some, alas, too late—that it can be as difficult to stop moving as to start. The very climate of the age, we may say, favoured the law's acceptance.

[54] In the second paragraph of the Third Day of the *Discourses Concerning Two New Sciences*, Galileo writes ([1638] 1974: 147): 'We shall divide this treatise into three parts. In the first part we consider that which relates to equable or uniform motion, in the second, we write of motion naturally accelerated; and in the third, of violent motion, or of projectiles' ([1638] 1974: 147–8). The title of a new section a few pages later is 'On Naturally Accelerated Motion' (ibid. 157).

[55] See Dawson's edition of *Hoyle's Games* (n.d.) on 'the origins of this noble game'. When, in *Anthony and Cleopatra*, II. v, Shakespeare gave Cleopatra the line 'Let's to billiards?', he was showing the groundlings that she was a member of the beau monde.

[56] See Pepys' *Diary*, 1 Jan. 1767.

4

Modelling, the Brownian Motion, and the Disunities of Physics

> Systematic unity (as mere idea) is only a *projected* unity, which one must regard not as given in itself, but only as a problem.
>
> Immanuel Kant[1]
>
> We cannot anticipate a 'final axiomatization'.
>
> Otto Neurath[2]

PREAMBLE

As a regulative ideal of physics, the importance of theoretical unification would be hard to deny. The achievements of Isaac Newton in the seventeenth century, and of James Clark Maxwell 200 years later, are celebrated precisely because each brought apparently diverse phenomena within the ambit of a single theory. Newton's mechanics accommodated both celestial and terrestrial phenomena; Maxwell's electromagnetic theory accommodated not only electrical and magnetic phenomena but optical phenomena as well.[3] Similarly, in the latter part of the twentieth century, practitioners of field theory partially realized the goal of a unified 'final theory', one which would accommodate the four 'fundamental forces' of nature. In the 1960s, electromagnetic forces and the weak nuclear force were unified in the electro-weak model proposed by Steven Weinberg and Abdus Salam. Subsequently, the electro-weak model and quantum chromodynamics (which deals with the strong nuclear force) were spliced together to create the 'Standard

---

[1] Kant [1781, A647] 1998: 593.   [2] Neurath [1938] 1955, 19.
[3] A more precise description of Maxwell's achievement is given by Margaret Morrison (1992: 118): '[W]hat was effected was the reduction of electromagnetism and optics to the mechanics of *one aether* rather than a reduction of optics to electromagnetism *simpliciter*' (emphases in the original).

Model', a theoretical advance which one historian has called 'one of the great achievements of the human intellect'.[4]

The Standard Model, however, contains nineteen parameters that must be put in 'by hand'.[5] It falls short of achieving d'Alembert's goal of 'reducing ... a large number of phenomena to a single one that can be regarded as their principle'.[6] And, at the time of writing, the force of gravity still resists incorporation; the final theory of the fundamental processes of nature remains a dream. But, even if this 'final theory' is vouchsafed to us, I shall suggest, the theoretical practices of physics will still exhibit disunity. Or, rather, they will exhibit disunities, for disunity comes in many forms. In this essay I examine five kinds of disunity, without claiming that my list is exhaustive. Each is illustrated by examples from modern physics. I describe these examples in realist terms, as a physicist would. At each stage of the discussion, the topic of unity and disunity is interwoven with another: the role of modelling in theoretical practice. In the latter part of the essay, I examine in some detail Einstein's use of modelling in his analysis of the Brownian motion, and draw a philosophical conclusion.

4.1. VARIETIES OF DISUNITY

4.1.1. Disunities of the First Kind: Scale

A molecule of H_2O is not wet, nor does an atom of iron display ferromagnetic behaviour. The macroscopic properties we associate with liquidity and ferromagnetism are inapplicable at the atomic scale. Molecules do not have a coefficient of viscosity; atoms do not have a magnetic susceptibility.

But, though the descriptive vocabularies of theories at the macroscopic and atomic scales may be disjoint, that does not by itself rule out the possibility of establishing relations between them. The reduction of thermodynamics to statistical mechanics, for example, relies on relations of just this kind.[7] In this reduction the pressure and temperature of a gas are represented as effects of the motions of the individual molecules that comprise it. The pressure is seen as the effect of the molecules bouncing elastically off the walls of the

---

[4] Schweber 1997: 172.   [5] See Kaku 1993: 14.

[6] Jean Le Rond d'Alembert [1751] 1963: 22.

[7] In *The Structure of Science* (1961: chap. 11), Ernest Nagel uses this as his paradigm example of the reduction of theories. The reduction is not without its problems (for a detailed discussion, see Sklar 1993, chap. 9).

vessel containing them, and is equated with the change of momentum the molecules suffer in these collisions per unit time, divided by the area of the walls. The temperature is regarded as a measure of the mean kinetic energy of the molecules. That said, the ideal gas is a very unusual system. It is the sum of its parts, composite, but not complex. The total kinetic energy of the molecules is just the arithmetical sum of their individual kinetic energies; likewise, the total change of momentum of the molecules in their collisions with a wall of their container is the arithmetical sum of their individual changes of momentum. Most importantly, the only interactions between the molecules themselves are elastic collisions, in which kinetic energy and momentum are both conserved.

In contrast, liquids and ferromagnets are genuinely complex systems, whose components interact in complicated ways. For this reason, macroscopic quantities like coefficient of viscosity and magnetic susceptibility cannot be expressed in terms of properties of individual molecules or atoms. In order to make the behaviour of systems like these intelligible, physicists typically choose one of two strategies. One is the use of narrative; they tell plausible stories.[8] The other is the use of tractable mathematical models to simulate the complex system's behaviour. To take one example of the first strategy: viscosity can be explained as a type of liquid friction which occurs when adjacent layers of a liquid move at different speeds.[9] When a molecule of the liquid migrates from a slower-moving layer to a faster-moving layer, it gets bumped from behind by the molecules in the faster-moving layer, so the story goes, and the effect of these collisions is to slow down the faster-moving layer. Conversely, when a molecule from the faster-moving layer migrates into a slower-moving layer, the effect of molecular collisions is to speed up the slower-moving layer. In both cases the net effect is to reduce the difference of speeds between the layers.

In the case of ferromagnetism, a very different strategy is provided by the *Ising model*.[10] Its specification is very simple. It consists of a regular array of *sites* in two- or three-dimensional space. The number of sites is very large. With each site is associated a variable s whose value can be either *up* or *down*;

---

[8] I illustrated the use of narrative in theoretical physics in section 2.2.4 of Essay 2, and I return to it in Essays 7 and 8.

[9] An example may occur when liquid passes through a cylindrical pipe and is retarded by the inner surface of the pipe. The physics of 'laminar flow' as opposed to turbulent was initiated by Navier and Stokes in the late nineteenth century, and further developed by Prandtl in the first decade of the twentieth. See Morrison 1999: 53–60 for discussion.

[10] Essay 6 is devoted to the Ising model and its role in theoretical practice.

an assignment of values to each of the sites in the model is referred to as an *arrangement*. With each pair of adjacent sites is associated an 'interaction energy', and its value for a given pair of sites depends on the arrangement. When the values of *s* for the two sites are antiparallel the 'energy' is *e*; when they are parallel it is −*e*. We may sum the 'interaction energies' for all pairs of adjacent sites to obtain $E(a)$, the total 'interaction energy' of the lattice under the arrangement *a*. The probability $p(a)$ that a particular arrangement will occur is also specified; it is defined by an equation (involving both $E(a)$ and a parameter *T*) that has the same form as Boltzmann's Law, the fundamental postulate of statistical thermodynamics.[11] The parameter *T* (which can take only positive values) plays the role of the 'absolute temperature' of the lattice; hence $p(a)$ is 'temperature-dependent'.

This completes the definition of the model. It is useful for two basic reasons. First, the two-dimensional version is (comparatively speaking) mathematically tractable, and computer simulations can be used to display the behaviour of both the two- and three-dimensional versions. Secondly, and most importantly, the model mimics the behaviour of a ferromagnetic material—in particular, its behaviour near the so-called *critical point*. If a ferromagnet is heated, its magnetic moment goes down with temperature at an increasing rate, and at the *critical temperature* (for iron, 770°C) its magnetic properties virtually disappear.[12] Critical-point behaviour analogous to this is also observed in physical systems as diverse as water, binary alloys, and superfluids. It turns out that each of these systems has a set of properties which are analogous to corresponding properties of a ferromagnet, and in each case the variation of these properties near a critical point is governed by a set of simple relations (the *scaling laws*); furthermore, for all such systems of a given dimension, these laws take exactly the same form. Thus, the behaviour of the Ising model sheds light on a wide variety of related phenomena.

This unification, however, serves only to emphasize the disunity under examination. It shows that behaviour at the scale where critical phenomena appear is effectively independent of the precise physical nature of the physical interactions at the atomic scale that bring it about. We may grant that these macroscopic phenomena would not appear, absent interactions at the

---

[11] The equation is: $pa = Z^{-1} \exp{-Ea/kT}$. *Z* is called the 'partition function' for the model, and its reciprocal Z^{-1} acts as a normalizing constant, to make sure that the probabilities add to one; *k* is Boltzmann's constant.

[12] In technical terms, it becomes a *paramagnetic* material, with a magnetic strength about a millionth of that of iron at room temperature.

atomic or molecular level, while noting that, as far as theoretical practice is concerned, the two levels are 'causally decoupled'. The interplay between unity and insularity is nicely caught in the title of an essay by Alastair Bruce and David Wallace (1989): 'Critical Point Phenomena: Universal Physics at Large Length Scales'.

The examples examined so far have all involved the relation between the properties of matter at the atomic/molecular scale and the macroscopic scale. To appreciate the range of scale involved, consider that a hydrogen atom has a diameter of about 10^{-10} metres, whereas the everyday scale at which we work starts around a millimetre, and our eyes can resolve distances as small as a few tenths of a millimetre.[13] The latter distances describe capacities of human beings; they indicate the limits of our dexterity and our organs of sight. Outside the biological realm, however, they are not physically significant. How important they are for the epistemology of science—which is, after all, the study of human knowledge—has been the topic of philosophical debate.[14] For present purposes, however, we may regard 10^{-10} m and 10^{-3} m simply as useful benchmarks. Both within the range they enclose and outside it there are physical processes and systems that are, so to speak, scale-specific, and whose theoretical treatment is correspondingly localized.

Within this range we find the physics of the nanoscale, sometimes called the *mesoscale*, because it is intermediate between the atomic and the macroscopic scales. A nanometer is 10^{-9} m, and the nanoscale runs from one to a few hundred nanometres. (In the lattice of carbon atoms that make up a diamond, nearest neighbours are 0.154 nanometres apart.) Recent technology has made possible the manufacture of nanoscale structures within which striking events occur. Although the structures are aggregates of atoms and molecules, they provide environments within which individual quantum events can be detected, the movement of single electrons can be controlled, and both electrical conductance and thermal conductance become quantized. One

[13] More precisely: consider two small points of light a distance s apart, both located at the near point of vision about 30 cm from the eye. Because of diffraction effects the eye will not form sharp images of these light sources on the retina. Instead, if s is less than about 0.1 mm, the two areas of the retina which receive light will overlap. Furthermore, even when the sources are just far enough apart for the illuminated areas to be separated, the receptors' rods in the retina are not sufficiently densely packed to register them as distinct areas. For obvious reasons, it is not enough for the two areas to be localized on separate, but adjacent, rods. The net result is that, unless s is more than 0.2 mm, the eye will see no separation between the sources.

[14] See, for example, Bas van Fraassen 1980: chap. 2, and Ian Hacking's paper 'Do We See through a Microscope' (Churchland and Hooker 1985: 132–52), to which van Fraassen responded (ibid. 297–300).

researcher claims, 'The past two decades have seen the elucidation of *entirely new, fundamental physical principles* that govern behaviour at the mesoscale' (my emphasis).[15] But, leaving aside the question of whether the nanoscale calls for a 'unique physics', as has been claimed, it is unquestionably the site of a major divergence in the physical sciences. For the nanoscale is the scale at which chemical bonding takes place between molecules. Here chemistry and biochemistry have established themselves as autonomous sciences, setting their own agendas and addressing them by their own methods.

Below the atomic scale the obvious example of a significant physical system is the atomic nucleus. The unit of length used in nuclear physics is the femtometre: $1 \text{ fm} = 10^{-15}$ m, and the effective radius of a stable nucleus is under 10 fm (i.e. less than one ten-thousandth of the diametre of the hydrogen atom).[16] The strong nuclear force holds the nucleus together despite the mutual repulsion of the positively charged protons in it. Within this environment, the strong force is about a thousand times as strong as the electromagnetic force, but has a very short range (of order 1 fm). Thus, it has no direct effect outside the immediate neighbourhood of the nucleus.[17] At the other end of the scale are the effects of very large masses on the space-time structure of the world. These effects are dealt with by the general theory of relativity, and a major reason why that theory languished between 1915 and 1960 is that they are hardly measurable except over astronomical distances. For example, the aim of one terrestrial experiment to confirm the theory (Pound and Rebka, 1960), was to establish the value of the measured red shift $\Delta v/v$ in the frequency of a γ-ray travelling vertically in the Earth's gravitational field. (v is its frequency in the absence of such a field.) The predicted value of the shift is 2.46×10^{-15}, less than three parts in a million billion.[18] Given the tiny range of nuclear forces, and the huge distances associated with general relativistic effects of any magnitude, it is hardly surprising that the theories that deal with them are ignored in the great bulk

[15] The claim is made on p. 50 of an article by Michael Roukes in the Sept. 2001 issue of *Scientific American* (Roukes 2001), which is devoted to nanoscience and nanotechnology. The phrase 'the unique physics' of the nanoscale appears in the subtitle of Roukes' article.

[16] The radial charge density for ^{208}Pb is illustrated by Jones (1987: 3). There is no sharp cut-off; the charge density drops off almost linearly between 5 and 8 fm. The distribution of matter is similar (ibid. 10).

[17] The weak force has a range of 100 fm, but is many orders of magnitude weaker than the strong force.

[18] The measured result, $2.57 \pm 0.26 \times 10^{-15}$, was remarkably close. See Weinberg 1972: 82–3.

of theoretical practice.[19] For the same reason, theoretical practice has largely treated them independently—apart from the attempts, mentioned earlier, to formulate a final theory that accommodates all four 'fundamental forces' of physics.

None the less, there are physical systems whose behaviour is governed by interactions between forces usually associated with widely different length scales. A neutron star, for instance, achieves equilibrium when the gravitational force that would otherwise make the star collapse is resisted by the 'degeneracy pressure' of the neutron gas within it.[20] While the gravitational forces are governed by general relativistic (rather than Newtonian) principles, the existence of a 'degeneracy pressure' is a quantum effect; the Pauli exclusion principle prevents any two neutrons in the same state from occupying the same region of space. A typical neutron star has a radius of around 10 km, a distance, one might think, that is many times greater than the range of quantum effects, but yet too small for general relativistic effects to be significant. Why then is either of these forces relevant to the star's behaviour? The answer lies in the extreme density of the star. A neutron star of radius 10 km has a mass roughly equal to that of the Sun. This has a double effect. On the one hand, the neutrons are packed very tightly together, so that quantum effects make themselves felt; on the other, the star is massive enough for general relativistic effects to come into play.

This 'double effect' brings me to the final point of this section of the essay. Up till now the only scale I have discussed is the length scale. But, as the example of the neutron star shows, the length scale is by no means the only scale that has physical significance. Others, like mass, temperature, energy, and time, are of comparable importance. And, as the phrases 'low-temperature physics', 'high-energy physics', and 'geological time' testify, they too give rise to disunities of the first kind.

4.1.2. Disunities of the Second Kind: Suppositions

A recurring leitmotiv here and in the two sections that follow is the correspondence between contemporary theoretical practice and some aspects of Descartes' philosophy of science. In the closing pages of the *Discourse*

---

[19] In contrast, effects of the kind dealt with by the special theory of relativity manifest themselves at the atomic scale.   [20] See Longair 1989: 133–40 and Will 1989: 23–5.

on the Method, Descartes returns to the problem of the gap between theory and phenomena. He tells us that, in his *Optics* and *Meteorology*, he has supplemented his fundamental principles with 'suppositions'. But, as he admits, the gap has not been completely bridged, since the suppositions have not been derived from fundamental principles. Instead, they have been obtained by arguing from effects to causes. That is, they have been put forward to account for specific phenomena, and then tested. As Descartes puts it ([1637] 2001: 61/76): '[A]s experience makes most of these effects very certain, the causes [i.e. the suppositions] from which I derive them serve not so much to prove these effects as to explain them; but, on the contrary, it is the causes which are proven by the effects.'[21]

By 1644, when Descartes published his *Principles of Philosophy*, 'suppositions' had become 'hypotheses'.[22] They are particularly valuable, Descartes tells us, in the investigation of 'imperceptible things':

Just as, when those who are accustomed to considering automata know the use of some machine and see some of its parts, they easily conjecture from this how the other parts which they do not see are made: so, from the perceptible effects and parts of natural bodies, I have attempted to investigate the nature of their causes and of their imperceptible parts.   ([1644] 1983a: IV. §203)

---

[21] Here, and in the sentence that follows, Descartes distinguishes between *proof* and *deduction*. Note that Olscamp translates 'deduire' by 'derive', a distinction which has appeared earlier in his paragraph as one between *proof* and *demonstration*. Within seventeenth-century philosophy, the primary philosophical sense of 'to prove', 'prouver', was to *test*, and the primary sense of 'to demonstrate', 'demonstrer', was to *deduce from first principles*, as in a geometrical proof. See Hughes 1990a: 177–9. When Descartes uses the words 'prouver' and 'demonstrer' early in the paragraph, he oscillates between them in a very artful manner, the effect of which is to suggest that, epistemologically, both procedures are on a par. He is addressing those who are shocked by some of the statements he makes, 'à cause que je les nomme des suppositions, et que je ne semble pas avoir envie de les prouver'. He responds, 'Il me semble que les raisons s'y entresuivent en telle forte que; comme les dernières sont demonstrées par les premieres, qui sont leurs causes, ces premieres le sont reciproquement par les dernières, qui sont leurs effets ...'. Note the absence of a verb in the penultimate clause. To my mind, Olscamp, like Robert Stoothoff in the Cambridge translation of Descartes' *Philosophical Writings* (1985), makes some wrong choices in translating these lines. First, he uses 'proven' to translate 'demonstrées', and so distorts what Descartes is saying; secondly, he inserts another 'proved' in the penultimate clause, and so masks Descartes' sleight of hand. I grant that it is not obvious how that clause may best be translated. In the references, the first two-digit number, ('61') in the present instance, gives the page in the Olscamp translation and the second two-digit number (here '76') guides the reader to the corresponding page on Vol. vi of the Adams and Tannery edition of Descartes' work.

[22] In fact, Olscamp translates the French 'supposition' in the *Optics* by 'hypothesis'. For a general discussion of the 'Method of Hypothesis' and its subsequent history up to the nineteenth century, see Laudau 1981.

Notice that the direction of the investigation is again from effects to causes.

Descartes' mention of 'automata' links his procedures to our own. The most obvious correspondence links them to the mechanisms whose role in scientific practice has been documented by Peter Machamer, Lindley Darden, and Carl Craven (2000). More abstractly, they are analogous to the phenomenological models that find employment in the twentieth-century version of the method of hypothesis.

Consider, for instance, the 'two-fluid' model of superconductivity. Superconductivity is now a well-known and widely applied phenomenon. At very low temperatures, the electrical resistance of some metals vanishes, so that no voltage is needed to keep a current flowing through them. It was first observed in 1911 by Kammerlingh Onnes, who found that at a transition temperature about $4°$ above absolute zero the resistance of mercury suddenly fell off by a factor of at least 10^5 (see Kubo and Nagamiya 1969: 186). The phenomenon has now been observed in many metals and alloys, and at considerably higher temperatures. In the two decades after its discovery, much experimental information became available about this phenomenon, but the first promising theoretical treatment of it was the two-fluid model proposed by Fritz and Heinz London in 1935 (ibid.). In this model, a distinction is made between two kinds of electric current, a 'normal current' and a 'superconducting current'. The 'London equations' describing the superconducting current are exactly those required to account for the Meissner effect—the observed effect that no magnetic field resides in a superconductor (ibid. 188–9).[23] Thus, the London brothers offer us a modern counterpart of a Cartesian hypothesis: a model whose characteristics have been tailored to fit the phenomena; which is to say, they have been established by arguing from effects to causes.

Oddly, both the normal and the superconducting currents are thought of as composed of electrons. Hence, the model leaves a major question unanswered: How can the same entities form two distinguishable fluids? To answer this, another model is needed. It was provided by Bardeen, Cooper, and Schrieffer in the mid-1950s. Whereas the only force between free electrons is the repulsion due to the Coulomb interaction between them, on the BCS model a pair of electrons can also 'use the lattice' to exert an attractive force on one another. The number of the 'Cooper pairs' suddenly increases at the transition temperature, and they comprise the superconducting current that the two-fluid London model requires.

[23] Other lessons from this episode are drawn by Mauricio Suarez 1999: 182–95.

For twenty years prior to the appearance of the BCS model, however, physicists had perforce to be content with a 'local' model for the phenomenon of superconductivity, within which electromagnetic theory was supplemented by a new relation between the superconducting current and a magnetic field—that is to say, by elements whose specification was independent of the fundamental equations of the theory. The practice of using such a model gives rise to a *disunity of the second kind.*

This is a purely formal, or, if you like, structural disunity. The use of these phenomenological models means that physics cannot be set out as a theory in the logician's sense. To echo Descartes, at any stage of physics we can expect to find a logical gap between the principles of our fundamental theories, 'so simple and so general', and many of the 'particular effects' to which these principles apply.

4.1.3. Disunities of the Third Kind: Principles

Never inclined to false modesty, in Part Six of his *Discourse on the Method*, Descartes outlines a programme designed to establish 'the general principles or first causes of all that is or can be in the world' ([1637] 2001: 52/63–4).[24] He hopes to derive these principles from nothing but 'certain seeds of truth which are naturally in our souls' (ibid.). He acknowledges, however, that he has run into a problem:

> But I must also confess that the power of nature is so ample and so vast, and these principles so simple and so general, that I almost never notice any particular effect such that I do not see right away that it can be derived from these principles in many different ways; and my greatest difficulty is usually to discover in which of these ways it depends on them.   (ibid.)

In short, there is a gap between theory and phenomena.

Three and a half centuries later, astronomers faced a problem of an analogous kind. It concerned the anomalous viscosity of accretion disks.[25] Many astronomical systems have the form of a disk; the most familiar

---

[24] In quoting the *Discourse on Method* and the *Optics* I use the translation by Paul Olscamp (Descartes [1637] 2001). Here and there I amend it to bring the translation closer to the original. Thus, in this quotation, I write: 'it depends on them', rather than 'it is derived', for 'il en depend'.

[25] I owe this example to the astronomer John Hawley, who presented it at a symposium on computer simulation held at the University of Virginia in the spring of 1999. The work he described appears in Balbus and Hawley 1998. In his presentation, he also referred the audience to the (1973) paper by Shakura and Sunyaev, to which I refer later.

examples are the solar system and the Milky Way. An *accretion disk* is formed when matter falling towards an astronomical body acquires angular momentum around it. For example, if a neutron star is coupled with a star similar to the Sun to form a binary system, an accretion disk will form when matter emitted as solar winds from the primary (Sun-like) star is attracted to the neutron star.[26] Because of the angular momentum associated with a binary system, this matter forms a gaseous accretion disk that rotates around the neutron star. The disk then acquires, as it were, a life of its own:

Once formed it is the disk itself that mediates continued accretion, and the physical processes that regulate mass inflow [into the neutron star] will generally be very different in character from those that may have triggered the initial infall.   (Balbus and Hawley 1998: 2)

If the material in the accretion disk is to spiral inwards towards the neutron star, then there must be a mechanism by which it loses some of its angular momentum. One possibility is that viscous forces within the gas bring about a transfer of angular momentum outwards through the disk. (From Kepler's laws, the smaller the radius of orbit of a particle, the higher its velocity; the difference of velocity between adjacent rings of the disk gives rise to viscous forces that tend to decrease the rate of rotation of the inner ring of the pair while increasing that of the outer ring. The net result will be a transfer of angular momentum away from the disk's centre.) The problem is that, given the accepted viscosity of the gas of the accretion disk, the viscous forces characteristic of laminar flow are not large enough to do the job. Indeed, 'the needed accretion rates are orders of magnitude higher than standard microsopic viscosities could provide' (ibid. 2b). Only if turbulence occurs and the viscosity becomes 'anomalous' can the angular momentum be redistributed to the extent required. In this way, the proposed solution of one problem has given rise to another. The search for a mechanism by which the accretion disk loses some of its angular momentum has become a search for the source of turbulence within the disk. But, where should this search begin? Accretion disks are complex systems, consisting of gas plasmas moving with supersonic orbital speeds. A complete analysis of their behaviour could involve physical principles from any or all of a slew of physical theories: general relativity, Newtonian mechanics, electrodynamics, hydrodynamics, magneto-hydrodynamics, plasma physics, and radiation transport among

---

[26] See Longair 1989: 139–44. The primary star of a binary system is the member of the system that emits radiation in the visible spectrum.

them.[27] The existence of such semi-autonomous branches of physics at the same scale constitutes a *disunity of the third kind.*

Descartes' problem and the astrophysicist's are structurally similar, but not identical. Descartes writes as though all physical effects can be deduced from the same cluster of principles, and that the philosopher's task is 'to discover in which [way] it depends on them'. In contrast, the astrophysicist assumes that the physical principles governing astronomical phenomena fall within the scope of one or more semi-autonomous branches of physics, and that his task is twofold: first, to identify the relevant branch or branches, and then to use their theoretical resources to provide a model of the phenomenon in question.

That was the task that Steven Balbus and John Hawley undertook. Their strategy was to combine analytic methods (the traditional methods of mathematical physics) with numerical simulations on a supercomputer. The computer is needed because the equations that describe the behaviour of accretion disks cannot be solved by analytic methods. Designedly, the numerical simulations involved major simplifications. Balbus and Hawley isolated possible sources of turbulence in the disk, modelling it first as a purely hydrodynamic system, and then as a magneto-hydrodynamic system. In each case, general relativistic effects were ignored. (Hence the treatment was 'not rigorously applicable to flow near neutron star surfaces' (ibid. 6b).) The results of the simulations were unequivocal:

[A] weak magnetic field of any topology leads directly and naturally to turbulence and outward angular momentum transport. ... [In contrast] not a shred of evidence was found for non-linear hydrodynamic instabilities in Keplerian flow.   (Balbus and Hawley 1998: 45a)

In this way Balbus and Hawley solved the problem of anomalous viscosity. But, their quasi-Cartesian methodology was not the only one that produced theoretical results. In retrospect, it is remarkable that, in the face of this disunity, considerable theoretical progress was made by simply ignoring the problem. The approach taken in an influential paper by Nikolai Shakura and Rashid Sunyaev (1973) completely bypasses the question of how disk turbulence comes about. Instead, the authors assume that turbulence exists, and invoke a new parameter 'α' that effectively acts as a constant of proportionality between properties associated with the transport of angular

---

[27] The list is taken from the handout to John Hawley's conference presentation (see n. 25).

momentum and the large-scale properties of the disk.[28] For a quarter of a century, the use of this parameter allowed astrophysicists to produce models of the way accretion disks behave, which had only a chimerical basis in theory.[29] It also gave rise, Hawley tells us, to 'a tendency to regard the source of angular momentum transport as known—it is alpha!' As he remarks, some models should be 'taken with a grain of salt'.[30]

4.1.4. Disunities of the Fourth Kind: Incompatibility

One component of Descartes' scientific methodology is not acknowledged in his *Discourse on Method*. It appears in the first and second discourses of the *Optics*. Descartes' announced aim in these discourses is limited: he will say nothing about the true nature of light, but instead will confine himself to explaining 'all those of its properties that experience acquaints us with ...' ([1637] 2001: 66/83). For this purpose, he says, 'I believe that it will suffice that I make use of two or three comparisons ...' (ibid.). These 'comparisons', for which he was sternly rebuked by Pierre Duhem,[31] are very different from the suppositions and hypotheses discussed in the *Discourse on Method* and the *Principles of Philosophy*. Whereas in the *Principles* (IV. §§205–6) hypotheses are held to be, at a minimum, 'morally certain', in the *Opticks* comparisons are just heuristically useful analogies. Three comparisons are drawn. The engravings that illustrate them depict a blind man, a wine-vat, and a gentleman in seventeenth-century costume whacking a tennis ball.[32]

---

[28] α is defined by the equation $W_{R\phi} = \alpha c_s^2$. $W_{R\phi}$ is the turbulent stress tensor which governs the transport of angular momentum within the gas, and c_s is the isothermal speed of sound in the gas which is dependent on bulk properties of the gas, like density and temperature. For definitions of these terms, see Balbus and Hawley 1998: 9–11.

[29] Ibid.12–13 offer the example of a disk model whose significant properties (mass, midplane temperature, thickness, etc.) are linked by a set of equations. The assumption of proportionality implicit in the definition of α guarantees that there are as many equations as variables, and, so to speak, locks the model in place.

[30] These remarks appeared in Hawley's handout to his conference presentation (see n. 25).

[31] Duhem speaks of 'these mechanical comparisons, whose logical validity would be exposed to many criticisms' (*PT*, 34), but refrains from saying that they are redolent of the 'ample and weak' English mind.

[32] Like the *Discourse on Method*, the *Optics* is included in vol. vi of the standard Adam–Tannery edition of Descartes' works (AT). In this edition, the same engraving of a blind man is used twice (AT 135, 136). The first use of the corresponding analogy, however, occurs fifty pages earlier (AT 84–5). The wine-vat appears at AT 86. The tennis player appears in three different engravings. One, representing the reflection of light, is reproduced twice (AT 93, 94); another, representing refraction, three times (AT 91, 97, 98); and the third, representing what we would call total internal reflection, once, at AT 99. A selection of these illustrations is given in the Olscamp translation.

In the first of these comparisons ([[1637] 2001: 67/84), the passage of light through a transparent medium is compared with the transmission of impulses along a stick, by means of which a blind man acquires knowledge of the path in front of him. Descartes suggests that, if we think in terms of this analogy, we will not find it strange that light can pass instantaneously from the Sun to the Earth. We will also be weaned from the belief that something material has to travel from an object to our eyes if we are to see it. In the second comparison (ibid. 69–70/86–7), the wine in the wine-vat is compared with the 'very subtle and very fluid matter' which fills the cosmos, and the grapes are compared to the 'coarse parts of air and other transparent bodies'. Just as the effect of the pressure exerted on the wine will give it a tendency to move in a given direction (even though no movement takes place), the effect of a luminous body is to induce a similar kind of action in the 'subtle and fluid matter' to which the wine is compared. Indeed, rays of light are 'nothing other than the lines along which this action tends'.

In the third comparison (ibid. 70–83/89–105), this action is assumed to obey the same laws as motion itself, and the behaviour of a tennis ball as it rebounds from a hard surface, or as it passes from one medium to another, is used to provide explanations of the reflection and refraction of light. The illustrations that accompany this comparison all share one striking feature. In each case, two models are displayed simultaneously. The left-hand side of the illustration is occupied by a naturalistic rendering of a tennis player serving a tennis ball, and the right-hand side by a geometrical figure featuring a circle whose diameter is larger than the tennis player is tall. Two elements of the illustrations do double duty, and so connect their naturalistic and geometrical parts. The path of the ball as it leaves the racquet is represented by a straight line to the centre of the circle, and other radii of the circle represent the ball's path after it is deflected; similarly, the horizontal line that runs from left to right across each illustration is both a diameter of the circle and a representation of a physical surface (in some cases the ground, in others, the surface of a pond). Mathematical modelling and physical analogy are here intertwined. Together they enabled Descartes to provide not only the first published statement but also the first theoretical explanation of the relation we now call 'Snell's Law'.[33]

---

[33] Whether Descartes established the law independently of Snell is not known. See Sabra 1981 100–3.

The use of comparisons did not die with Descartes. Chapter 3 of G. A. Jones's *The Properties of Nuclei* (1987) begins:

In the previous chapters, simple nuclear models have been introduced, namely the liquid drop and Fermi gas models [...]. This chapter will discuss three models: the shell model, the collective model, and the optical model.[34]

Among these, the liquid-drop model, the shell model, and the optical model are analogue models, twentieth-century versions of Descartes' comparisons.

As one might assume, the liquid-drop model represents the nucleus as a drop of (incompressible) 'nuclear liquid'. Like a drop of water, the nucleus can be said to have a 'surface tension'. If the nuclear drop becomes deformed, this surface tension will act to bring it back to a spherical shape, and the ensuing vibrations may be severe enough to cause the drop to split into two. In this way, the model offers a theoretical account of nuclear fission, the process by which a heavy nucleus splits into two smaller nuclei of roughly equal size.

The shell model deals with a very different issue: Why is it that some nuclei are markedly more stable (i.e. less prone to decay) than others? In particular, why is stability found in nuclei in which the number of protons or the number of neutrons is one of the so-called 'magic numbers' (2, 8, 20, 28, 50, 82, or 126)? In this model the nucleus is compared to the atom, and the stability of certain nuclei to the chemical stability of inert gases (helium, neon, argon, etc.). In this representation, just as the electrons within the atom occupy certain shells, so do the nucleons (protons and neutrons) within the nucleus. In the atom, each shell can only accommodate a certain number of electrons. In an atom of an inert gas at the ground state all occupied shells are filled, starting with the innermost (K) shell and moving outwards. In the shell model of the nucleus, the nucleus is pictured as a potential well. That is to say, a nucleon at the centre of the nucleus will have a very low energy, and to leave the nucleus—to climb out of the well—it will have to gain energy. The nucleons are confined to specific energy levels in the well, and, in accordance with the Pauli exclusion principle, a given energy level can be occupied by at

---

[34] Jones's list is not exhaustive. C. M. H. Smith (1966: 620) lists eight 'theoretical frameworks' for treating the nucleus. They include not only four of the five models that Jones lists (only the Fermi gas model is absent) but also the α-particle model, the compound nucleon model, direct nuclear reactions, and the statistical model. Of the two books, Smith's is the more informative about the models and the uses to which they are put—or rather, *were* put, since the book was published in 1966. With the exception of the liquid-drop model, all the models considered assume that nuclei are made up of protons and neutrons, that the number of protons in nuclei of a particular element is equal to the atomic number of the element, and that isotopes of an element differ by virtue of the numbers of neutrons they contain.

most one nucleon. The 'shells' of the model each contain a certain number of energy levels. There are two sets of shells, one for protons and one for neutrons. A proton (neutron) shell contains all the proton (neutron) energy levels within a certain range of energy. Thus, the number of energy levels in a given shell determines the number of protons (or neutrons) that will fill it. If, starting from the lowest energy shell, the number of protons (or neutrons) in a particular nucleus is sufficient to fill a set of shells without remainder, that number is a 'magic number'.

An apparent difficulty for the model is that it pays no attention to collisions between nucleons, despite the fact that they are tightly packed within the nucleus. It was resolved by Victor Weisskopf, who pointed out that any collision that occurred within the nucleus would leave it unaffected. Since all the system's permissible energy states are occupied, and by Pauli's principle (again) only one nucleon can occupy a particular state, the net effect of any scattering could only be to make nucleons 'change places', as it were. But, in this case, the distinction between the initial and final states of the system would be a distinction without a difference. It would be as though the collision had never happened.

The optical model provides a theoretical account of what happens when a high-energy neutron enters a nucleus from outside. If the energy of the incoming neutron is considerably greater than the energies of the nucleons within the nucleus, the considerations advanced in the last paragraph no longer apply. In fact, the higher the neutron's energy, the greater will be the likelihood of it undergoing a collision in the nucleus—more precisely, the shorter will be its mean free path. The nucleus is compared to 'a "cloudy crystal ball"', partially transparent to the incoming particle and partially absorptive' (Smith 1966: 633). '[It] can be regarded as a homogeneous medium of a complex refractive index' (ibid. 632).

The collective model is a generalization of the shell model. Whereas in the latter model the shells in the nucleus have spherical symmetry, in the former a 'collective' motion of the nucleons in an unfilled shell can take place, giving rise to vibrations or rotations of the nucleus as a whole. The model was first proposed to account for nuclear excitation energies that were much higher than the values given by the shell model, but it was also used to explain a phenomenon observed when a beam of high energy particles triggers nuclear fission in a target nucleus. The fragments from some nuclei are preferentially emitted at right angles to the incident beam, while the distribution of fragments from others is nearly isotropic.

The last of Jones's five models, the Fermi gas model, is very simple. A Fermi gas is a collection of particles governed by the laws of quantum mechanics—in particular, by the Pauli principle. The model represents the nucleus as two non-interacting Fermi gases, one composed of protons, and the other of neutrons. Despite its simplicity, the model can be used to suggest which nuclei of a specific atomic number will have the least stored energy and, for that reason, will be the most stable. Its definition is so general, however, that the model can have very little predictive or explanatory power.[35]

The liquid-drop model differs from the others in important ways. For the moment, I will set it aside and see how the others are related to each other. All four of them represent the nucleus as a quantum mechanical system. By virtue of the distinction drawn between proton shells and neutron shells, the shell model is a special case of the Fermi gas model. The collective model, in turn, is a generalization of the shell model. Whereas the latter represents the nucleus as spherical, and deals most successfully with nuclei with closed shells, the former treats the violations of spherical symmetry induced by the collective motion of 'loose' nucleons in the outermost, partially filled shell. The shell model and the collective model supplement that of the Fermi gas model with significant auxiliary assumptions. They yield models that conform to a fundamental theory, but contain elements whose specifications are independent of that theory. In other words, they introduce disunities of the third kind. Though each of them employs different assumptions, they may, nevertheless, be thought of as complementary modifications of a basic prototype, designed to deal with specific aspects of nuclear behaviour.

Both treat the nucleus as a potential well, as does the optical model. The difference between the liquid model and, say, the shell model introduces a disunity of a different kind—the fourth kind. An obvious difference is that the liquid-drop model represents the nucleus as a classical system rather than a quantum mechanical system, but that is only part of the story. The chief difference resides in the incompatibility between the narratives that the two models supply. John Ziman's account of the situation is worth quoting:

Like Alice's Red Queen, who could 'believe six impossible things before breakfast', it may even be necessary to accept logically contradictions of models of the same system. In the theory of atomic nuclei, the essence of the 'liquid drop model' is that

---

[35] As noted above (n. 34), Smith does not include the Fermi gas model in his list of 'theoretical frameworks'. Jones describes it as a quantum version of the liquid-drop model, but he does not show how it can replicate that model's main achievement, the explanation of nuclear fission.

the protons and neutrons within the nucleus are packed so densely together and repel each other so strongly that they form something like an incompressible fluid. This model is beautifully validated by phenomena such as nuclear fission, where the drop becomes unstable and divides in two. On the other hand, to explain the stability and radioactive properties of a large number of nuclei, the so-called 'shell model' has to assume that each proton or neutron can move with almost perfect freedom inside the whole volume of the nucleus, almost as if the other particles did not exist. This model, also, gives quantitative results in excellent agreement with experiment. It is easy enough to assert optimistically that these two models are merely idealizations or specialized aspects of a much more complex but unified theory which will eventually yield the correct description of the phenomena in all cases: but until, by some theoretical *tour de force*, this unification is actually achieved, the nuclear physicist is faced with manifestly contradictory models, each of which is so scientifically sound that it has earned a Nobel Prize for its authors and proponents. (Ziman 1978: 37)

In a footnote to this passage Ziman reminds us that, when we talk of aspects of theoretical practice, we must (at least mentally) index them by date.[36] He writes:

This was certainly the case in the 1950s. It may be that the theoretical unification has now been formulated. I could easily find out by asking the experts, or by reading a few review papers. But the subject is highly technical, and not very interesting in itself. The status of a theory changes with time, and must not be judged on its 'final' state.   (Ziman 1978: 37 n. 40)

4.2. EINSTEIN, BROWNIAN MOTION, AND INCONSISTENCY

4.2.1. Einstein and Brownian Motion

At this point I turn aside—or appear to turn aside—from my main argument to the debate in the late nineteenth and early twentieth centuries between the energeticists, on the one hand, and proponents of the 'atomic hypothesis', on the other. The papers on Brownian motion that Einstein published between 1905 and 1908 were instrumental in the debate's resolution, and I will examine the first of them in some detail. I have two reasons for doing so.

---

[36] On indexing by sub-discipline and date, see the Preamble to Essay 2. Note that Ziman wrote the passage quoted and the footnote that accompanies it in 1978.

The first is that the 1905 paper offers a remarkable case study of the use of modelling in theoretical practice; the second is that it will lead us towards disunities of the fifth (and final) kind.

'Brownian motion' is the name given to the random movements of tiny particles suspended in a liquid. Though these movements are small, they are large enough to be observed through a microscope.[37] The molecular explanation of the motion is that in any short period of time a great number of the molecules of the liquid will bounce off the suspended particle and tend to propel it in some direction or other. The movements we observe are due, not to individual collisions between the suspended particle and particularly energetic molecules of the liquid, but to statistical variations in the momentum imparted to the particle in different directions as time goes on; at one time the particle may receive more blows pushing it from left to right than from right to left, for example, and, a bit later, a greater number pushing it up rather than down.

It is only with hindsight that Einstein's 1905 paper can be unequivocally described as concerned with Brownian motion. As he announces in the title, the motion he sets out to analyse is 'the movement of small particles ... demanded by the molecular-kinetic theory of heat'. In other words, his aim is not to explain a motion that has already been observed, but to predict and describe a motion that might prove observable. Whether the observed and the predicted motions will turn out to be one and the same remains to be seen; Einstein writes ([1905] 1956: 1): 'It is possible that the movements to be discussed here are identical with the so-called "Brownian molecular motion"; however, the information available to me regarding the latter is so lacking in precision, that I can form no judgement in the matter.'

Neither here nor elsewhere does Einstein point out how important it is for his project that the two motions be identified. If they were independent of each other, then the phenomena he predicts could always be masked by Brownian movement, and so could never provide evidential support for the molecular theory of heat he favours.[38]

As it is, the evidence they provide is remarkably indirect. I have described the cause of Brownian motion rather vaguely, as due to 'statistical variations

---

[37] A brief history of work on Brownian motion prior to 1905 is given by R. Furth in Einstein 1956: 86–8 n. 1.

[38] This may explain why in 1906 Einstein accepted so readily Siedentopf's assurance that 'he and other physicists ... had been convinced by direct observation [*sic*] that the so-called Brownian motion is caused by the irregular thermal movements of the molecules of the liquid' (Einstein [1906b] 1956: 19).

in the momentum imparted by the particle in different directions as time goes on'. Einstein is even vaguer. 'The irregular movement of the particles', he says (ibid. 11), is 'produced by the thermal molecular movement'. A page later (ibid. 12), he says again that the 'irregular movements' of the suspended particles 'arise from thermal molecular movement'. These two references to 'thermal molecular movements' are the nearest Einstein comes to a causal account of the random motion of the suspended particles.

Instead, his strategy is to apply to suspensions the account of solutions that he put forward in his doctoral dissertation.[39] On this account, when a solute is taken into solution, the molecules of the solute behave like small particles moving around within the solvent. They effectively constitute a 'gas' of molecules, in that their behaviour can be described in terms of the models provided by the kinetic theory of gases. As we saw in section 4.1.1, these models represent a gas as a collection of molecules in random motion; the (absolute) temperature T of the gas is seen as proportional to the mean kinetic energy of the molecules, and the pressure P that the gas exerts on the walls of the container as the force per unit area exerted by the molecules in bouncing off them. According to Einstein's theory of solutions, it is no coincidence that van 't Hoff's law for dilute solutions has exactly the same form as the ideal gas law:[40]

$$PV = RTz \tag{3.2.1}.$$

When this law is applied to solutions, P becomes the osmotic pressure resulting from the presence of a mass z of solute in the solution.

On this analysis, the molecules of the solute behave like particles suspended in the surrounding solvent. Conversely, Einstein suggests, a suspension of particles in a liquid may be expected to behave like a solution. In particular, the suspended particles may be expected to exert an 'osmotic pressure' due to thermal motions.

According to this theory, a dissolved molecule is differentiated from a suspended body *solely* by its dimensions, and it is not apparent why a number of suspended particles should not produce the same osmotic pressure as the same number of molecules (Einstein [1905] 1956: 3).

---

[39] The dissertation ([1906a] 1956) was completed only eleven days before the 1905 paper was submitted for publication, and was not itself published until 1906.

[40] In fact, van 't Hoff defined *ideal solutions* as 'solutions which are diluted to such an extent that they are comparable to ideal gases' (Pais 1982: 87).

Observe the role played by modelling in these two analyses. To begin with (*a*) a dissolved substance is modelled as a suspension of molecules; and (*b*) a gas is modelled as a collection of molecules in random motion. Both these subject–model relationships are then inverted. That is to say (*d′*) a suspension of molecules is represented as a dissolved substance in a solution; and (*b′*) a collection of molecules in random motion is represented as a gas. In the doctoral dissertation (*a*) is then coupled with (*b′*): a dissolved substance is modelled as a suspension of molecules, and this in turn is modelled as a gas. In this way, Einstein obtains an expression for the osmotic pressure P. Then, by (*d′*), he models a suspension as a solution, and interprets P as an 'effective pressure' exerted by the suspended particles.

In an important respect, however, the suspended particles and the postulated solute molecules both differ from the molecules of an ideal gas. They move within a viscous fluid, whereas the molecules of an ideal gas do not. Einstein uses this fact in deriving an expression for the diffusion coefficient D of the particles (or solute molecules), which specifies the rate at which particles diffuse from regions of high concentration to regions of low concentration.[41] He obtains,

$$D = RT/6\pi \eta aN \qquad\qquad (3.2.2).$$

Here η is the coefficient of viscosity of the fluid, and a the radius of the particles (assumed spherical). The equation is very revealing. R and T appear because the particles are being modelled as a gas obeying the ideal gas law; Avogadro's number N appears because, in accordance with the kinetic theory of gases, the gas is being modelled as a Newtonian system of particles; and η and a appear because these particles are represented as moving through a viscous fluid. Its mission accomplished, the hypothetical pressure P exerted by the suspended particles has disappeared from view.

Einstein now considers the thermal motions which, on the kinetic theory, bring about this diffusion. As I mentioned earlier, there is no analysis of what causes these motions. All Einstein assumes is (i) that there is a specific probability function $\Phi(\Delta) + d$ that the particle will move a distance between Δ and $\Delta + d$ in a certain time interval τ, (ii) that this function is non-zero only when Δ is small, and (iii) that it is symmetric about $\Delta = 0$. Einstein

---

[41] The derivation is wonderfully cunning; for an analysis of Einstein's argument, see Pais 1982: 90–8.

confines himself to motion is one dimension. From assumptions (i)–(iii) he obtains the root mean square value λ of the distance moved by a particle in time *t*:

$$\lambda = \sqrt{(2Dt)} \qquad (3.2.3)$$

where D is defined in terms of Φ and shown to be equal to the diffusion coefficient. Hence (3.2.2) and (3.2.3) together give,

$$`\lambda = \sqrt{[2t.\ RT/6\pi\eta aN]} \qquad (3.2.4)$$

from which,

$$N = RTt/3\pi\eta a\lambda^2 \qquad (3.2.5).$$

Assuming a value for N of 6×10^{23} molecules per mole, Einstein calculates that, if $t = 1$ minute, then $\lambda \sim 6\mu$, a distance large enough to be observed through a microscope.

As we have seen, in this analysis Einstein first models a solution as a suspension (move (*a*)), and then models a suspension as a solution (move (*a'*)). Each of these moves effects a reversal of the observable/unobservable distinction.[42] In the case of a solution, the osmotic pressure is measurable (a 1 per cent sugar solution typically exerts an osmotic pressure of about $^2/_3$ atmospheres; see Pais 1982: 87), but the molecules and their motions are not. In contrast, the 'effective pressure' attributed to particles in a suspension is too small to measure. In equation (3.2.1) z is measured in moles (or gm molecular wt). On the molecular theory, therefore, $z = n/N$, where n is the number of molecules in the volume V. Equation (3.2.1) then becomes:

$$P = (RT/V) \times (n/N) \qquad (3.2.1^*).$$

When this equation is applied to the suspension, n/N is tiny, and so is P. (Thus, no measurement of P can provide a direct challenge to classical thermodynamics, according to which no such pressure exists.[43]) On the other

---

[42] *Contra* van Fraassen 1980, I follow Hacking 1985 in classing things we see with a microscope more accurately, with many kinds of microscope as observable.

[43] Einstein [1905] 1956: 2–3 notes the conflict between the two theories, but does not mention that the difference is too small to detect.

hand, unlike the molecules in a solution, the particles in a suspension are themselves observable, and so are their motions.

Indeed, in the same decade in which Einstein's paper appeared, Jean Perrin and his co-workers found ways to measure the relevant quantities in equation (3.2.5), and thereby obtained a value for Avogadro's number. These results, along with others, established the reality of molecules in the eyes of everyone except Ernst Mach.[44]

4.2.2. Disunities of the Fifth Kind: Contradiction

Einstein's analysis, however, contains a paradoxical feature, to which I have so far drawn no attention. Throughout his discussion, he uses a model which is almost invisible to the reader; it remains, quite literally, in the background. Recall that the molecules of a solute and the particles of a suspension are both treated as systems of particles moving through a viscous fluid. Equation (3.2.2), for example, contains the factor $6\pi\eta a$ because, by Stokes's Law, which is used in its derivation, the drag F on a sphere moving with speed v through a fluid is given by $F = 6\pi\eta av$. But, ordinary hydrodynamics, within which this law appears, models a fluid as a continuous and homogeneous medium. And, in section 2 of the paper in which Einstein shows that 'the existence of an osmotic pressure can be deduced from the molecular-kinetic theory of Heat', he assumes that 'the liquid is homogenous and *exerts no force on the* particles' (Einstein [1905] 1956: 9–10; emphasis added).

This, on the face of it, is bizarre. On the one hand, Einstein announces that an experimental confirmation of the result he predicts would be critical for the molecular-kinetic theory of heat, on which matter is treated as particulate. On the other, in making these predictions he treats matter (at least in the liquid state) as continuous and homogeneous. Just as a disunity of the fourth kind arises when incompatible models are used to treat different phenomena, a disunity of the fifth kind occurs when the assumptions employed within a single analysis are mutually at odds. Furthermore, he is quite explicit about what he is doing. In a 1908 review of his work on solutions and suspensions, he writes:

But when the dissolved molecule can be looked on as approximately a sphere, which is large compared with a molecule of the solvent, we may ascertain the

---

[44] In a one-line paragraph, Pais records (1982: 103): 'Mach died in 1916, unconvinced.'

frictional resistance of the solute molecule according to the methods of ordinary hydrodynamics, which does not take account of the molecular constitution of the liquid.   (Einstein [1908] 1956: 73)

Reader, if you find this peculiar, then you are probably not a physicist. A good physicist has a finely tuned sense of when to use one model and when another. Like Einstein, she or he can be untroubled by a disunity of the fifth kind.

5

Models and Representation

We form for ourselves images [*innere Scheinbilder*] or symbols [*Symbole*] of external objects; and the form which we give them is such that the necessary consequents of the images in thought are always the images of the necessary consequents in nature of the things pictured.

Heinrich Hertz[1]

Few terms are used in popular and scientific discourse more promiscuously than 'model'.

Nelson Goodman.[2]

PREAMBLE

A major philosophical insight, recognized by Heinrich Hertz and Pierre Duhem at the turn of the nineteenth century, and recovered eighty years later by proponents of the semantic view of theories, is that the statements of physics are not, strictly speaking, statements about the physical world.[3] They are statements about theoretical constructs. If the theory is satisfactory, then these constructs stand in a particular relation to the world. To flesh out these claims, we need to say more about what kinds of constructs are involved, and what relation is postulated between them and the physical world. To call these constructs 'models', and the relation 'representation', does not get us very far. Scientific models, as I pointed out in Essay 4, are many and various. If our philosophical account of theorizing is to be in terms of models, then we need both to recognize this diversity and to identify whatever common elements exist within it. One characteristic—perhaps the only characteristic—that all theoretical models have in common is that they

---

[1] Heinrich Hertz [1894] 1956: 1.   [2] Goodman 1968: 171.
[3] Frederick Suppe, an early bird, wrote (1977a: 224): 'What the theory actually characterizes is not the phenomena in its intended scope, but rather idealized replicas of these phenomena.' Idealization is the least of it, as I will show in due course.

provide representations of parts of the world, or of the world as we describe it. But, the concept of representation is as slippery as that of a model. On the one hand, not all representations of the world are theoretical models; Vermeer's 'View of Delft' is a case in point. On the other, the representations used in physics are not, in any obvious sense, all of one kind. What, we may ask, does the representation of DNA as an arrangement of rods and plates have in common with the representation of the motion of a falling body by the equation $s = gt^2/2$? Apart, of course, from being the sort of representation that a model provides.

5.1. GALILEO'S DIAGRAMS

Galileo's writings offer a good place to start. Much of the Third Day of his *Discourses and Mathematical Demonstrations Concerning Two New Sciences* is given over to kinematics, specifically to 'Naturally Accelerated Motions'. In Proposition I, Theorem I of this part of the work Galileo relates the distance travelled in a given time by a uniformly accelerating object starting from rest with that travelled in the same time by an object moving with a uniform speed. He concludes that the two distances will be equal, provided that the final speed of the accelerating object is twice the uniform speed of the other. To demonstrate this proposition (i.e. to prove it), Galileo uses a simple geometrical diagram (Fig. 5.1). He writes:

Let line *AB* represent the time in which the space *CD* is traversed by a moveable in uniformly accelerated movement from rest at *C*. Let *EB*, drawn in any way upon *AB*, represent the maximum and final degree of speed increased in the instants of the time *AB*. All the lines reaching *AE* from single points of the line *AB* and drawn parallel to *BE* will represent the increasing degrees of speed after the instant *A*. Next I bisect *BE* at *F*, and I draw *FG* and *AG* parallel to *BA* and *BF*, the parallelogram *AGFB* will [thus] be constructed, equal to the triangle *AEB*, its side *GF* bisecting *AE* at *I*. ([1638] 1974: 165)

After a short discussion, he concludes:

[I]t appears that there are as just as many momenta of speed consumed in the accelerated motion according to the increasing parallels of triangle *AEB*, as in the equable motion according to the parallels of the parallelogram *GB* [*sic*]. For the deficit of momenta in the first half of the accelerated motion (the momenta represented by the parallels on the triangle *AGI* falling short) is made up for by the momenta represented by the parallels of triangle *IEF*.   (ibid. 167)

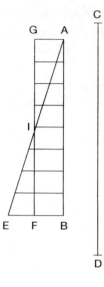

Fig. 5.1. Naturally Accelerated Motion (1).

Galileo's analysis is couched explicitly in terms of representations. A time interval is represented by the length of a vertical line *AB*; an object's speed at any instant is represented by the length of a horizontal line parallel to *BE*.[4] Effectively, after drawing attention to the equality of the two areas *AEB* and *AGFB*, Galileo takes this equality to represent the equality of the distances travelled.[5]

Another diagram (Fig. 5.2) accompanies the discussion of Corollary I to Proposition II. The corollary states:

[I]f there are any number of equal times, say *AC, CI, IO*, taken successively from the first instant or beginning of [a uniformly accelerated] motion, then the spaces traversed in these times will be to one another as are the odd numbers from unity, that is, as 1, 3, 5 ...   (ibid. 167)

---

[4] Though the parallel lines representing speeds are drawn in his diagrams as horizontal lines, Galileo does not insist on this; they are drawn, he says 'in any way' upon the time axis. That is why he refers to *ABFG* as a 'parallelogram'. Thus, the representation is given only as much geometrical structure as his problem needs.

[5] 'Effectively', because Galileo never gives a rigorous justification for representing a distance by an area. In fact, the only geometric or kinematic magnitude he claims to represent by an area is an area. In place of areas, his argument invokes 'aggregates of all parallels contained in [e.g.] the triangle *AEB*', each parallel representing 'the increasing degree of speed after the instant speed' ([1638] 1974: 165). In this section, I will continue to behave anachronistically.

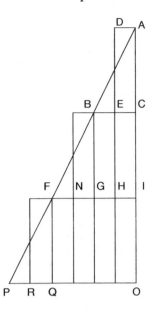

Fig. 5.2. Naturally Accelerated Motion (2)

In the diagram, alongside the line segments *AC*, *CI*, and *IO* that represent the first three time intervals, we find, respectively, one, three, and five rectangles, all congruent to each other.[6]

Galileo's strategy, here and in many other sections of the *Two New Sciences*, is to take a given problem, in this case a problem in kinematics, and represent it geometrically. He then reads off from the geometrical representation the solution to the problem. In brief, he reaches his answer by changing the question; a problem in kinematics becomes a problem in geometry.

This simple example suggests a very general account of theoretical representation. I call it the *DDI account* (see Fig. 5.3) On this account the use of a model in physics involves three components: *denotation, demonstration*, and *interpretation*. The three components appear very clearly in Galileo's corollary: time intervals are denoted by distances along a vertical axis, speeds by lengths of horizontal lines. Demonstration then takes place entirely within the model. Elementary geometry shows us that the three areas, *ABC*, *CBFI*, and *IFPO* are in the ratio $1:3:5$. We can then interpret this

---

[6] For brevity, I have changed the wording of the Corollary. Between the diagrams I have labelled Figure 5.1 and Figure 5.2, Galileo introduces another, where the lines *AB*, *BC*, etc. refer to different magnitudes.

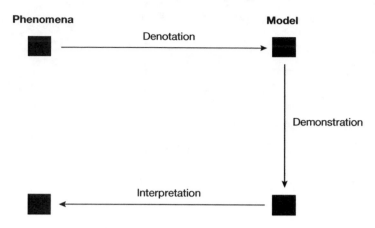

Fig. 5.3. A schematic diagram of the DDI account of theoretical representation.

geometrical result kinematically: a ratio of areas can be interpreted as a ratio of distances.

But, this is not the whole story. Simplicio (Galileo's Aristotelian commentator) remarks:

I am [now] able to see why the matter must proceed in this way, once the definition of uniformly accelerated motion has been postulated and accepted. But I am still doubtful whether this is the acceleration employed by nature in the motion of her falling heavy bodies.   (ibid. 169)

Simplicio does not offer an alternative account of 'the acceleration employed by nature', but asks Salviati (representing Galileo) to describe the experiments that led Galileo to his conclusions. Salviati's response is interesting:

[I]n order to be assured that the acceleration of heavy bodies falling naturally does follow the ratio expounded above, I have often made the test [*prova*] in the following manner, and in [Galileo's] company.

   In a wooden beam ... a channel was rabbeted in along the narrowest dimension, a little over an inch wide and made very straight; so that this would be clean and smooth, there was glued within it a piece of vellum, as smoothed and cleaned as possible. In this there was made to descend a very hard bronze ball, well rounded and polished ...   (ibid.)

Nature, it seems, needed some assistance. And, however much smoothing and cleaning took place, we may still doubt that the acceleration of the bronze ball had the perfect uniformity of the hypotenuse of an Euclidean triangle. Even so simple an example of representation as Galileo's has its complexities.[7]

The DDI account handles these complexities with ease, as I will show. But, to establish that, I need to say more about the three components that, on this account, are characteristic of theoretical representation.

5.2. DENOTATION

Let me forestall possible misunderstandings. I am not arguing that denotation, demonstration, and interpretation constitute a set of acts individually necessary and jointly sufficient for an act of theoretical representation to take place. I am making the more modest suggestion that, if we examine a theoretical model with these three activities in mind, we shall achieve some insight into the kind of representation that it provides. Furthermore, we shall rarely be led to assert things that are false.

We shall not, for instance, be led to say that theoretical models resemble, or are similar to, their subjects in certain respects and to specified degrees.[8] There is, of course, a type of scientific model of which this is true. It is true, for instance, of the model used in the mid-eighteenth century by Joseph Smeaton to investigate the relative merits of overshot and undershot waterwheels. This was a miniature waterwheel similar to working waterwheels in a precise geometrical sense. It would be a defect of the DDI account if it did not extend to material models like these; nevertheless, in their similarity to their subjects they are the exception, not the rule.[9]

---

[7] The diagrams also diverge from the experiments in a subtler way. The salient variables in the experiments are time and length, and the dependency of one variable on the other is reversed when we move from the experiments to the diagrams. In one of the experiments, Salviati tells us ([1638] 1974: 170): '[T]he spaces were found to be to each other as the squares of the times', but it would be more accurate to say that the times were found to be to one another as the square roots of the spaces. These times, we are told, were measured by a water clock, 'a large pail filled with water ... which had a slender tube affixed to its bottom, through which a narrow thread of water ran [into] a little beaker. ... The little amounts of water collected in this way were weighed ... on a delicate balance' (ibid.). These 'little amounts of water' were those collected during the time taken for the ball to traverse previously marked distances. Galileo examined 'now the time for the whole length [of the channel], the time of one half, or that of two thirds, or three quarters ...' (ibid.). In the diagrams, by contrast, the time axis, as we may call it, is marked off in equal times, and a geometrical construction is performed to obtain the areas that represent distances traversed in those times.   [8] As does Ronald Giere, for example; see Giere 1985: 90.

[9] For his work on waterwheels Smeaton was awarded the 1759 Copley Prize by the Royal Society. I am indebted for this example to my colleague Davis Baird, who compares Smeaton's

To take a more typical example, we may model an actual pendulum, a weight hanging by a cord, as an ideal pendulum. We may even be tempted to say that in each case the relation between the pendulum's length and its period of oscillation is approximately the same, and in that respect they are similar to each other. But, the ideal pendulum has no length, and there is no period of time in which it completes an oscillation. It is an abstract object, similar to material pendulums in no obvious sense. More illuminating, albeit less fashionable, would be the suggestion that material pendulums participate in its form. Instead, I propose that we attend to Nelson Goodman's dictum (1968: 5) that 'denotation is the core of representation and is independent of resemblance'. Following Goodman, I take a model of a physical system to 'be a symbol for it, stand for it, refer to it' (ibid.). Just as a vertical line in one of Galileo's diagrams denoted a time interval, elements of a scientific model denote elements of its subject; in Duhem's words (*PT*, 20), theoretical terms 'bear to the latter only the relation of sign to the thing signified'.

Goodman was concerned with pictorial representation; the book I have just quoted is *Languages of Art*. But, fifteen years before that work appeared, Stephen Toulmin had proposed a view of scientific theory that placed a similar emphasis on denotation. A theory, said Toulmin ([1953] 1960: chap. 4), is like a map. The aptness of the analogy is not surprising, since maps and models have much in common. Their affinity comes out very clearly in everyone's favourite example, Henry C. Beck's map of the London Underground, which displays an intricate network of differently coloured lines. Each line segment is either parallel to the edge of the map, or perpendicular to it, or at 45° to it; coloured red or blue, black or brown, pink or yellow as the case may be, it can hardly be said to resemble the section of track which it denotes. It is natural to think of this map as providing a model of the system it represents. Conversely, Watson and Crick's rod-and-plate model of DNA can be thought of as a map in three-dimensional space.

The positive analogy between a map and a physical theory is considerable.[10] For present purposes, however, it is more instructive to look at the negative

---

model to the theoretical analysis put forward by Antoine Parent in 1704, thus (2004: 31): 'Parent's theory and Smeaton's model both provided representations of waterwheels. Parent's theory had built-in assumptions about efficiency and work. With Smeaton's model, they were literally built in. Smeaton, unencumbered by misleading theory and well informed by his practical experience with waterwheels, was better to represent them in a material model than Parent was in equations.'

[10] See 'World Maps and Pictures' (John Ziman 1975: chap. 4) and Toulmin [1953] 1960: chap. 4.

analogy between them, the places where they differ, rather than where they correspond. The first disanalogy is that a map refers only to an existing particular, whereas in general a physical theory represents a whole class of systems, some actual, others merely possible. Given this generality of scope, perhaps the view that a scientific theory can denote a multitude of systems needs to be argued for, rather than just asserted. Observe, first, that some theories do represent specific particulars. Cosmology does so, by definition. There are also physical theories that apply only to specific types. An obvious example is Bohr's theory (or model) of the hydrogen atom.[11] I will assume without argument that our concept of denotation allows us to denote a type. (Think of the pictures in a *flora* or an illustrated dictionary.) I will call a theory like Bohr's a *local theory*, and the model it defines a *local model*. It deals with a clearly defined type of physical system. In contrast, a *global theory* like classical mechanics or quantum theory deals with a heterogeneous collection of physical systems. In a trivial sense, all the systems dealt with by, say, quantum mechanics can indeed be described as belonging to one type; each is a token of the type 'system-whose-behaviour-can-be-modelled-by-quantum-mechanics'. More to the point, however, each individual system of that type can be represented by a particular model definable in terms of the theory, specifically by its Hamiltonian. In short, each system can be represented as a particular kind of *quantum system*, and it is at that level of specificity that denotation comes into play.

A global theory defines, not a particular model, but a class of models that, if the theory is fertile, continually expands. In Newtonian mechanics these models involve different configurations of masses and forces, all governed by Newton's laws of motion. Book I of Newton's *Principia*, for example, is devoted to the motion of theoretical systems consisting of non-interacting masses moving in a centripetal force-field; section I of Book II to the motions of masses in a resisting medium.[12] Basic texts on quantum mechanics, on the other hand, start from the simple energy configurations that make up Hamiltonians—infinite potential wells, simple harmonic oscillators, and Coulomb potentials—and from them deduces the permissible wave-functions of systems in these configurations.

---

[11] Max Jammer, in *The Conceptual Development of Quantum Mechanics*, gives a brief history of proposals concerning the structure of the hydrogen atom (1966: 70–4). Curiously, he credits Kelvin, Helmholtz, Bjerknes, Perrin, Thomson, Nagaoka, and Nicholson with *models* of the atom, Bohr alone with a *theory* of it.   [12] See section 3.5 of Essay 3.

In either case, when we apply the theory we either settle for one of the standard models or order a custom-built one. Essentially, a global theory is a set of instructions for building a great variety of local models. In any application of the theory it is a local model that represents an individual physical system, or type of system. There is thus no significant difference between the representations supplied by local and global theories; neither presents a challenge to the maxim: No representation without denotation.

5.3. DEMONSTRATION

The second disanalogy is that theories, unlike maps, are always *representations-as*, representations of the kind exemplified by Joshua Reynolds' painting of Mrs Siddons as the Muse of Tragedy. In this painting, Reynolds invites us to think of his primary subject, Mrs Siddons, in terms of another, the Muse of Tragedy, and to allow the connotations of this secondary subject to guide our perception of Mrs Siddons. Similarly, the wave theory of light represents light as a wave motion. It invites us to think of optical phenomena in terms of the propagation of waves, and so anticipate and explain the behaviour of light.

This aspect of theoretical representation is not confined to analogical models. Perhaps surprisingly, it is also characteristic of abstract mathematical representations, and is made explicit when we talk, for instance, of representing a plasma as a classical system, or as a quantum system. A mathematical representation should not be thought of simply as an idealization or an abstraction. Like an analogical representation, it presents us with a secondary subject that has, so to speak, a life of its own. In other words, the representation has an internal dynamic whose effects we can examine. From the behaviour of the model we can draw hypothetical conclusions about the world over and above the data we started from.

Toulmin and Ziman make the same claim for maps.[13] They point out, quite rightly, that the number of geographical relationships that can be read off a map by far exceeds the number of triangulations and distance measurements needed to draw it. But, although maps have this property, they are not representations-as; they introduce no secondary subject in terms of which we are invited to think of the landscape. Instead, while the map itself,

---

[13] See Toulmin [1953] 1960: 110–11 and Ziman 1978: 82–5.

a two-dimensional Euclidean surface, is capable of encoding this additional information, that is the limit of its capacities. The result is that, though the additional facts that maps supply may be new, they are never novel, as Toulmin points out ([1953]1960: 107). They may make us exclaim, 'Fancy that!' but never, 'How can that be?' They are not like the bright spot in the centre of the shadow of a small disk whose existence was entailed by Fresnel's wave theory of light, or the peculiar joint probabilities of space-separated pairs of events predicted by quantum mechanics. Both these predictions were of novel facts. Like many novelties, they engendered disquiet. It took Arago's experiments to convince some of his contemporaries that the first was not manifestly absurd.[14] As for the second, even after Aspect's experiments, the question, 'How can that be?' is still with us.

Of course, not all theories lead to novel conclusions. But, the same internal dynamic that permits the prediction of novel facts will also allow the prediction of everyday ones. Equally, it will allow us to confirm that the theory tallies with what we already know. Its function is epistemological. To be predictive, a science must provide representations that have a dynamic of this kind built into them; that is one of the reasons why mathematical models are the norm in physics. Their internal dynamic is supplied, at least in part, by the deductive resources of the mathematics they employ. Even so, mathematical models are not the only ones possible. Material models also possess the requisite dynamic. If we want to apply the wave theory of light to two-slit interference, for example, we can do one of two things. We can either model the phenomenon mathematically, or we can get out a ripple tank and model light as a wave motion in the literal sense. In either case we will find that the distance between interference fringes varies inversely both with the separation of the sources and with the frequency of the waves.

The same result appears whether we use the mathematical or the material model. The internal dynamic of the mathematical model is supplied by a mixture of geometry and algebra; that of the material model by the natural processes involved in the propagation of water waves. The internal dynamic of a computer simulation of this phenomenon would be something else again. But, all these modes of representation share a common feature. Each of them acts as an epistemic engine, generating answers to questions we put to it. I choose the term 'demonstration' for this activity in order to play upon its diachronic ambiguity. Whereas in the seventeenth century geometrical

---

[14] See 'Appendix 14: Poisson's Spot' in Jed Buchwald 1989: 373–6.

theorems were said to be 'demonstrated', nowadays we demonstrate physical phenomena in the laboratory. Mathematical models enable us to demonstrate results in the first sense, material models in the second.

5.4. INTERPRETATION AND THE NESTING OF MODELS

The results we demonstrate with a ripple tank are, in the first instance, results about ripples. From them we draw conclusions about the behaviour of light. Galileo's diagram shown in Figure 5.2 displays three sets of rectangles. From the number of rectangles in each set we infer the ratios of three distances. In each case the conclusions demonstrated within the model have to be interpreted in terms of its subject. *Interpretation*, the third movement of the triad, *denotation, demonstration, interpretation*, yields the predictions of the model. Only after interpretation can we see whether theoretical conclusions correspond to the phenomena, and hence whether the theory is empirically adequate.

Heinrich Hertz, whose account of theories is an ancestor of the representational view,[15] thought of theories as providing 'images' ('*innere Scheinbilder*'). In theorizing, he writes ([1894] 1956: 1): 'We form for ourselves images or symbols of external objects. ... The images which we here speak of are our conception of things. They are in conformity with things in one important respect, namely in satisfying [a specific] requirement.' The requirement is that of empirical adequacy: we require that '[T]he necessary consequents of the images in thought are always the images of the necessary consequents in nature of the things pictured'.

To compare vocabularies, what I call 'denotation' is for Hertz the relation between images and external objects. The necessity Hertz attaches to the 'necessary consequents of the images in thought' is a theoretical necessity, associated on my account with demonstration, and defined by the dynamic of the model. In contrast, the necessity he attaches to the necessary consequents of the things pictured is physical necessity.[16] On the DDI account, interpretation is a function that takes us from what we have

---

[15] I introduced and elucidated the phrase 'the representational view of theories' in section 3.2.2 of Essay 3.

[16] Here I am glossing, not endorsing, Hertz's account. Whether we have, or need, a coherent notion of physical necessity is a subject for debate.

demonstrated (the necessary consequents of the images) back into the world of things. Hertz's requirement stipulates that these theoretical consequents must themselves be images; specifically that they must denote the 'necessary consequents of the things pictured'. The requirement of empirical adequacy is thus the requirement that interpretation is the inverse of denotation.

This is an elegant, albeit obvious, result. None the less, its elegance cannot disguise the fact that Hertz's account of theoretical representation, and the DDI account as I have so far presented it, are in an important respect misleading. I have hinted as much in my brief discussion of the local models governed by global theories. In that discussion, I pointed out that in order to apply a foundational theory to a particular physical system we need to model that system in terms of the configurations amenable to theoretical treatment. This procedure may itself require considerable ingenuity, even if the physical system is comparatively simple. Notoriously, if we model a 3-body system within Newtonian celestial mechanics, analytic solutions of the relevant equations are unobtainable. Newton's strategy in treating the solar system is, first, to represent it as an abstract system of non-interacting bodies in a centripetal field, and subsequently to adjust this model to allow for perturbations due to interactions between selected pairs of bodies. Similarly, to give a quantum mechanical account of even so elementary a system as the hydrogen atom, we start by modelling it in terms of the Coulomb potential between an electron and a proton; only later do we take account of the spin of the electron by adding so-called 'radiative corrections' to allow for the Lamb shift.[17] We fall short of guaranteeing that 'the necessary consequents of the images in thought are always the images of the necessary consequents in nature of the things pictured'.

On their own, complexities of this sort do not require us to modify the basic DDI account of representation. They simply show that, even in a mathematical theory, demonstration cannot be equated with pure mathematical deduction; of comparable importance are approximative methods and perturbation techniques.[18] But, they also suggest that the resources available for demonstration may influence, and indeed dictate, the form of the theoretical model. They suggest that before we can apply the equations of the theory we will need, in Cartwright's words (1983: 133), a *prepared description* of the phenomena.

---

[17] For a discussion of this example, see Cartwright 1983: 137–8.
[18] See section 2.2.7 on Deduction in Essay 2.

A simple example of a prepared description is provided by Bohm and Pines in their quartet of papers, 'A Collective Description of Electron Interactions' (1950–3), which I presented as an example of theoretical practice in Essay 2. Bohm and Pines inherited what was already an accepted model of a metal in the solid state: a lattice of positive ions immersed in a gas of electrons (the 'conduction electrons'). The central papers of the quartet, BP II and BP III, each contained a theoretical analysis to show that in certain conditions the conduction electrons would execute collective longitudinal oscillations. The accepted model was already idealized, in that any impurities in the metal were ignored. In addition, Bohm and Pines disregarded the discrete nature of the ions; the effect of the ions was represented by a uniform positive background. Not until the idealization step had weeded out impurities, and the abstraction step had smoothed out the discrete nature of the positive ions, was the model ready to be described in strictly mathematical terms. That a conceptual gap existed between the model and its theoretical description is easy to see: the model was represented in two different ways, by classical physics in BP II, and quantum mechanics in BP III. In both papers, the passage to the theoretical level involved the representation of a representation.

Until Cartwright's work in the early 1980s, philosophers of physics had paid scant attention to this hierarchy of theoretical representations. Yet, as the example of the BP quartet shows, no account of the application of theory based on a simple dichotomy between observational and theoretical statements, or between external objects and our images of them, will do justice to theoretical practice. In contrast, the DDI account readily accommodates a hierarchy of this kind. For, the model used at an early stage of theorizing (model$_1$) can itself be the subject of another model at a deeper level (model$_2$). Take the example of BP III, where model$_1$ is the electron-gas-against-positive-background model, and model$_2$ the system redescribed in the language of quantum mechanics. As Fig. 5.4 shows, the whole three-step process associated with model$_2$ (denotation$_2$-demonstration$_2$-interpretation$_2$) effectively acts as the demonstration step for model$_1$; it takes us from images supplied by model$_1$ to their theoretical consequences. The theory in question provides demonstration$_2$. Denotation$_2$ and interpretation$_2$ conform with the theoretical practices that link the two models.

The question we started with was this: What kind of representation does a scientific model provide? The answer, in summary form, is that a scientific model provides a representation-as: it represents a primary subject in terms of a secondary subject, the model. The internal dynamic of the model

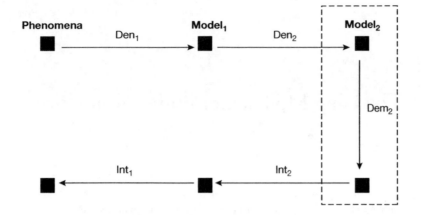

Fig. 5.4. Hierarchies of representation on the DDI account of modelling.

makes possible the demonstration of new, and sometimes novel, conclusions. Representations of this kind may be nested one within another, so that one model provides the internal dynamic for another. Designedly skeletal, this account needs to be supplemented on a case-by-case basis to reveal, within individual examples, the strategies of theory entry, the techniques of demonstration, and the practices, theoretical and experimental, that link theoretical prediction with experimental test.

6

The Ising Model, Computer Simulation, and Universal Physics

So the crucial change of emphasis of the last twenty or thirty years that distinguishes the new era from the old one is that when we look at the theory of condensed matter nowadays we inevitably talk about a 'model'.

Michael Fisher[1]

PREAMBLE

It is a curious fact that the index of *The New Physics* (Davies 1989), an anthology of eighteen substantial essays on recent developments in physics, contains only one entry on the topics of computers and computer simulation. Curious, because the computer is an indispensable tool in contemporary research. To different degrees, its advent has changed, not merely the way individual problems are addressed, but the sort of enterprise in which theorists engage. Consider, for example, chaos theory. Although the ideas underlying the theory were first explored by Poincaré at the turn of the century, their development had to await the arrival of the computer.[2] In *The New Physics*, the striking pictures of fractal structures that illustrate the essay on chaos theory (Ford 1989) are, of course, computer generated. Yet, perhaps because it runs counter to the mythology of theoretical practice, that fact is mentioned neither in the captions that accompany the pictures nor elsewhere in the text. The indispensable has become invisible.

The solitary entry on computers in the index takes us to the essay, 'Critical Point Phenomena: Universal Physics at Large Length Scales', by Alastair Bruce and David Wallace, and, within that essay, to a description of the so-called *Ising model* and the computer simulation of its behaviour.

---

[1] Fisher 1983: 47.   [2] See Poincaré 1908.

The model is at the same time very simple and very remarkable. It is used to gain insight into phenomena associated with a diverse group of physical systems—so diverse, in fact, that the branch of physics that explores what is common to them all is called 'universal physics'.

This essay has two sections. In section 6.1, I set out the relations between the phenomena, the Ising model, and the general theory of critical-point behaviour, and then outline the role played by computer simulations of the Ising model's behaviour. In section 6.2, I show how the Ising model in particular, and computer simulations in general, can be accommodated within a philosophical account of theoretical representation.

6.1. CRITICAL-POINT PHENOMENA AND THE ISING MODEL

6.1.1. The Phenomena

Various apparently dissimilar physical systems—magnets, liquids, binary alloys—exhibit radical changes in their properties at some critical temperature.

(a) Above the Curie temperature, T_C (770°C), a specimen of iron will exhibit paramagnetic rather than ferromagnetic behaviour; that is to say, above T_C it can be only feebly magnetized, below T_C its magnetic susceptibility is very high.[3]

(b) At the boiling point of H_2O, two phases, liquid and vapour, can coexist. The boiling point increases smoothly with pressure until the critical point is reached ($p_C = 218$ atmospheres, $T_C = 374$°C). At this point, the two phases cannot be distinguished; to quote Thomas Andrews, lecturing on similar behaviour in CO_2 at this point: 'If anyone should ask whether it is now in the gaseous or liquid state, the question does not, I believe, admit of a positive reply.'[4]

(c) Within a narrow range of temperatures around its critical temperature, a colourless fluid may exhibit critical opalescence, 'a peculiar appearance of moving or flickering striae throughout its entire extent'.[5] By 1900, this effect had been observed in a large number of fluids; subsequently Smulakowsky

---

[3] The elementary treatment of ferromagnetism given in Lee 1963: chap. 7 is still useful.
[4] Andrews 1869, quoted Domb 1996: 10.   [5] Andrews 1869.

Theoretical Practices of Physics

and Einstein attributed it to fluctuations in the refractive index of the fluid caused by rapid local changes in its density.

(d) In the 1920s and 30s, X-ray diffraction experiments on various binary alloys (e.g. copper-gold, copper-zinc) indicated that a transition from order to disorder within an alloy's crystal lattice could occur at a critical temperature which was well below the alloy's melting point.

The list is far from exhaustive—transitions to and from a superconducting or a superfluid phase are obvious other examples[6]—but the four above indicate the diversity of critical-point phenomena.[7]

6.1.2. The Ising Model

An Ising model is an abstract model with a very simple structure. It consists of a regular array of points, or *sites*, in geometrical space. Bruce and Wallace consider a square lattice, a two-dimensional array like the set of points where the lines on a sheet of ordinary graph paper intersect each other; the dimensionality of the lattice, however, is not stipulated in the specification. The number N of lattice sites is very large. With each site I is associated a variable s_i, whose values are $+1$ and -1. An assignment of values of s_i to the sites of the lattice is called an *arrangement*, *a*, and we write '$s_i(a)$' for the number assigned to the site I under that arrangement. The sum $\Sigma_i s_i(a)$ then gives us the difference between the number of sites assigned $+1$ and -1 by *a*. We may express this as a fraction of the total number N of sites by writing:

$$M_a = (1/N)\Sigma_i s_i(a) \tag{6.1}.$$

If we think of an arrangement in which each site in the lattice is assigned the same value of s_i as *maximally ordered*, then it is clear that M_a gives us the degree of order of an arrangement and the sign of the predominant value of s_i. For the two maximally ordered arrangements, $M_a = \pm 1$, for any arrangement which assigns $+1$ to exactly half the lattice sites, $M_a = 0$.

So far, the specification of the model has no specific physical content. Now, however, two expressly physical concepts are introduced. First, with

---

[6] See, e.g., Pfeuty and Toulouse 1977: 151–2 and 166–7, respectively.

[7] The four I list here are discussed in Domb 1996, a book which is simultaneously a textbook of critical-point physics and an internal history of its development. It is very comprehensive, if somewhat indigestible. As an introduction to critical-point physics, I recommend Fisher 1983.

each adjacent pair of sites, $<j, k>$, is associated a number $-Js_j s_k$, where J is some positive constant. This quantity is thought of as an interaction energy associated with that pair of sites. It is a function of s_j and s_k, negative if s_j and s_k have the same sign, positive if they have opposite signs. The total interaction energy E_a, the arrangement a, is the sum of these pairwise interaction energies:

$$E_a = \Sigma_{<j,k>} - Js_j(a)s_k(a) \qquad (6.2)$$

where $<j,k>$ ranges over all pairs of adjacent sites. Clearly, E_a is at a minimum in a maximally ordered arrangement, where $s_j = s_k$ for all adjacent pairs $<j, k>$.

Secondly, the lattice is taken to be at a particular temperature, which is independent of the interaction energy. Between them, the interaction energy E_a and the absolute temperature T determine the probability of p_a of a given arrangement; it is given by Boltzmann's law, the fundamental postulate of statistical thermodynamics:

$$p_a = Z^{-1} \exp(-E_a/kT) \qquad (6.3)$$

(k is Boltzmann's constant); Z is called the *partition function* for the model; here its reciprocal Z^{-1} acts as a normalizing constant, to make sure that the probabilities sum to one. To use the terminology of Willard Gibbs, equation 6.3 expresses the fact that the set of arrangements forms a *canonical ensemble*. The probability of a given arrangement decreases with E_a, and hence with the number of adjacent pairs of opposite sign. As T approaches absolute zero, disordered arrangements have a vanishing probability, and it becomes virtually certain that the system will be in one or other of the fully ordered minimal energy arrangements (which one depends on the system's history). On the other hand, an increase in temperature will flatten out this dependence of p_a on E_a. As a result, since there are many more strongly disordered arrangements than there are strongly ordered arrangements, at high temperatures there will be a very high probability of nearly maximal disorder, in which roughly equal numbers of sites will be assigned positive and negative values of s_j.

The probable degree of order, so to say, is measured by the *order parameter*, M. This parameter is obtained by weighting the values of M_a for each arrangement a by the probability p_a of that arrangement:

$$M = \Sigma_a p_a M_a \qquad (6.4).$$

Since we are keeping J (and hence E_a) constant, and summing over all possible arrangements, M is a function of the temperature, and is at a maximum at absolute zero. Between the extremes of low and high temperature, the Ising model exhibits the analogy of critical-point behaviour. There is a critical region, a range of values of T in which the value of M drops dramatically, and also a critical temperature, T_C, above which M is effectively zero.

Another significant temperature-dependent parameter is the *correlation length*. It is defined indirectly. Although the couplings between adjacent sites act to correlate the values of s_i at sites several spacings apart, thermal agitations tend to randomize them. The result is that the correlation coefficient $\Gamma(r)$ for the lattice falls off exponentially with distance r.[8] We write, $\Gamma(r) = \exp(-r/\xi)$. Since ξ in this equation has the dimensions of length, we may regard the equation as an implicit definition of the *correlation length*.[9] Effectively this parameter provides a measure of the maximum size of locally ordered islands, as it were, within the lattice as a whole.[10] At low temperatures, where the lattice is highly ordered, these islands are locally ordered regions within which the value of s_i is opposite to the predominant one; thus, if M_a for a typical arrangement is close to -1, they are regions where s_i is uniformly positive. At high temperatures, the relevant islands exist within a sea of general disorder, and can be of either sign. At both high and low temperatures ξ is small. Near the critical temperature, however, these islands can be very large; indeed, in the infinite Ising model, as T approaches T_C from either direction, ξ tends to infinity. It is worth emphasizing that ξ gives a measure of the *maximum* size of locally ordered regions. As will appear in section 6.1.7, the emergence of critical-point phenomena depends crucially on the fact that, near the critical temperature, islands of all sizes up to the correlation length coexist, and participate in the behaviour of the model.

---

[8] A correlation function for a pair of sites at separation r is defined as follows. For a given arrangement a we denote the values of s_i at the sites $s_0(a), s_r(a)$. Mean values of s_0 and the product $s_0 s_r$ defined by: $<s_0> =_{\mathrm{df}} Z^{-1} \Sigma_a \exp(-H/kT) s_0$; $<s_0 s_r> =_{\mathrm{df}} Z^{-1} \Sigma_a \exp(-H/kT) s_0 s_r$. ($H$ is the total energy of the lattice.) Since, if s_0 and s_r are uncorrelated, we would expect $<s_0 s_r> = <s_0><s_r>$, we define the correlation function $\Gamma(r)$ by $\Gamma(r) =_{\mathrm{df}} <s_0 s_r> - <s_0><s_r>$. When $H = E_a$, $<s_0> = 0 = <s_r>$, and so $\Gamma(r) = <s_0 s_r>$. The more general recipe given here allows for assigned or added energy terms, like those associated with an external magnetic field.

[9] Formally, $\xi = -r/\ln\Gamma(r)$. Since $\Gamma(r)$ is never greater than one, its logarithm is never positive.

[10] Strictly, fluctuations allow locally ordered regions to appear and disappear. The correlation length ξ may more properly be thought of as a measure of the size of those locally ordered regions that persist for a certain length of time. We may also see a correlation time that measures the mean lifetime of such regions (see Pfeuty and Toulouse 1977: 5).

6.1.3. Interpretations of the Ising Model

Although Bruce and Wallace focus on the two-dimensional Ising model for expository reasons, notably the ease with which computer simulations of its behaviour can be displayed, there are, in fact, various kinds of physical systems for which it provides a model. Pfeuty and Toulouse (1977: 4) show properly Gallic tastes in listing 'films, adsorbed phases, solids made up of weakly coupled planes (like *mille-feuille* pastry)'.[11] But, as I mentioned earlier, nothing in the specification of the model is peculiar to the two-dimensional case. In this section, I will go back to the phenomena I listed earlier, and see how the abstract three-dimensional Ising model can be interpreted in terms of them.

(a) Ising himself thought of the site variable in his model as the direction of the magnetic moment of an elementary magnet. As he wrote later:

At the time [the early 1920s] Stern and Gerlach were working in the same institute [The Institute for Physical Chemistry, in Hamburg] on their famous experiment on space quantization. The ideas we had at that time were that atoms or molecules of magnets had magnetic dipoles and that these dipoles had a limited number of orientations.[12]

The two values of s_i in the abstract Ising model are interpreted as the two possible orientations, 'up' or 'down', of these dipoles, and the coupling between neighbouring dipoles is such that less energy is stored when their moments are parallel than when they are anti-parallel. The order parameter, M, for the lattice, is interpreted as the magnetization of the ferromagnetic specimen, and the correlation length as a measure of the size of magnetic domains, regions of uniform dipole orientation.

(b) and (c) A model of liquid-vapour mixture near the critical point that is isomorphic to the Ising model was introduced by Cernuschi and Eyring in 1939, and the term 'lattice gas' first appeared in papers by Yang and Lee in 1952.[13] In the lattice gas, the Ising model is adjusted so that the two values of s_i are 1 and 0. The sites themselves are thought of as three-dimensional cells, and the value, 1 or 0, assigned to a site, is taken to denote the presence or absence of a molecule in the corresponding cell. For an adjacent pair of

---

[11] See also Thouless 1989 on condensed matter physics in fewer than three dimensions.

[12] Ernest Ising, letter to Stephen Brush (n.d.), quoted Brush 1967: 885–6. On Stern and Gerlach, see Jammer 1966: 133–4.   [13] See Domb 1996: 199–202 and Brush 1967: 890.

cells, $<j,k>$, $-Js_js_k$ takes the value $-J$ when $s_j = s_k = 1$, and zero otherwise. In other words, there is no interaction between an empty cell and any of its neighbours. The order parameter now depends on the fraction of cells occupied—that is, the mean density of the liquid-vapour mixture, and the correlation length on the size of droplets within the vapour. Local fluctuations in the density give rise to local variations in refractive index, and hence to critical opalescence.

(d) A binary alloy is a mixture of atoms of two metals, A and B. More realistic here than a simple cubic Ising model, in which all lattice sites are equivalent, would be a lattice that reflected the crystal structure of a particular alloy (e.g. a specific brass alloy). Very often such lattices can be decomposed into two equivalent sub-lattices. For example, if we take a cubic lattice, whose sites can be labelled $a_1, a_2, \ldots a_k, \ldots$, and introduce into the centre of each cube another site b_j, then the set $\{b_j\}$ of sites introduced will constitute another lattice congruent with the first. Which is to say that a body-centred cubic lattice can be decomposed into two equivalent cubic lattices interlocked with each other. The value of s_i that an arrangement assigns to each site I of this composite lattice represents the species of the atom at that site. An arrangement is perfectly ordered when each a-site is occupied by an atom of species A, and each b-site by an atom of species B, and maximally disordered when the atoms are randomly assigned to sites. We may define a parameter S_α analogous to the function M_a defined for the Ising model, and hence derive for the composite lattice an order parameter that depends, not on pairwise couplings between neighbouring atoms, but on the long-range regularity present in the system as a whole.[14] As in the standard Ising model, the greater the degree of order in the lattice, the less the amount of energy stored in it; it was by analogy with the idea of long-range order in alloys that Landau (1937) introduced the term 'order parameter' into the theory of critical phenomena in general.[15]

---

[14] I have assumed that there are as many A-atoms as a-sites, and B-atoms as b-sites. For this special case, let p_a denote the proportion of a-sites occupied by A-atoms under the arrangement α. Then we may write: $S_\alpha =_q [p_\alpha - p_{random}]/[p_{perfect} - p_{random}] = [p_\alpha - \frac{1}{2}]/[1 - \frac{1}{2}] = 2p_\alpha$ To confirm that the behaviour of S_α mimics that of M_a in the Ising model described earlier, observe what occurs (i) when all A-atoms are on a-sites, (ii) when A-atoms are randomly distributed over the lattice, and (iii) when all A-atoms are on b-sites. Bragg and Williams 1934: 701–3 generalize the definition to accommodate the general, and more realistic case, when there are fewer A-atoms than a-sites.

[15] See Domb 1996: 18. Note, however, that many authors (e.g., Amit 1984 and Domb himself) still use 'magnetization' rather than 'order parameter' as a generic term (see 6.1.6 below).

Notice that if we take a simple cubic Ising model and label alternate sites along each axis as *a*-sites and *b*-sites, then it becomes a composite lattice of the general type I have just described. The set of *a*-sites and the set of *b*-sites both form cubic lattices whose site spacing is twice that of the original lattice. As before, we let the value of s_i represent the species of atom at site I. Using an obvious notation, we allow the interaction energies between neighbouring sites to take the values J_{AA}, J_{AB}, J_{BB}. If $J_{AB} < J_{AA}$ and $J_{AB} < J_{BB}$, then arrangements of maximum order (with A-atoms on *a*-sites and B-atoms on *b*-sites, or conversely) become arrangements of minimum energy. The ordered regions whose size is given by the correlation length are now regions within which the value of s_i alternates from one site to the next.

Simple cubic crystal lattices are rare, and it might appear that, to be useful in any individual case, the Ising model would have to be modified to match the crystal structure of the alloy in question. As we shall see, however, there is an important sense in which the simple cubic Ising model adequately represents them all.

6.1.4. A Historical Note

The success of the Ising model could scarcely have been foreseen when it was first proposed. The digest of Ising's 1925 paper that appeared in that year's volume of *Science Abstracts* ran as follows:

A Contribution to the Theory of Ferromagnetism. E. Ising (1925: 253). An attempt to modify Weiss' theory of ferromagnetism by consideration of the thermal behaviour of a linear distribution of elementary magnets which (in opposition to Weiss) have no molecular field but only a non-magnetic action between neighbouring elements. It is shown that such a model possesses no ferromagnetic properties, a conclusion extending to a three-dimensional field.   (W.V.M.)

In other words, the model fails to leave the ground. Proposed as a model of ferromagnetism, it 'possesses no ferromagnetic properties'. Small wonder, we may think, that Ising's paper was cited only twice in the next ten years, on both occasions as a negative result. It marked one possible avenue of research as a blind alley; furthermore, it did so at a time when clear progress was being made in other directions.

In the first place, in 1928, Heisenberg proposed an explanation of ferro-magnetism in terms of the newly triumphant quantum theory. His paper was, incidentally, one of the places in which Ising's work was cited. Heisenberg wrote: 'Ising succeeded in showing that also the assumption of directed sufficiently great forces between two neighbouring atoms of a chain is not sufficient to explain ferromagnetism.'[16] In the second place, the *mean-field* approach to critical-point phenomena was proving successful. At the turn of the twentieth century, Pierre Curie and Pierre Weiss had explored an analogy between the behaviour of fluids and of magnets. Weiss had taken up van der Waals' earlier suggestion that the intermolecular attractions within a gas could be thought of as producing a negative 'internal pressure',[17] and had proposed that the molecular interactions within a magnet might, in similar fashion, be thought of as producing an *added* magnetic field. He wrote, 'We may give $[H_{int}]$ the name *internal field* to mark the analogy with the internal pressure of van der Waals' (Domb 1996: 13). The van der Waals theory of liquid-vapour transitions, and the Weiss theory of ferromagnetism (the very theory that Ising tells us he tried to modify) were examples of *mean-field theories*—so called because in each case an internal field was invoked to approximate the effect of whatever microprocesses gave rise to the phenomena. Other examples were the Curie–Weiss theory of anti-ferromagnetism, and the Bragg–Williams theory of order-disorder transitions in alloys.[18] The latter was put forward in 1934, the authors noting that 'the general conclusion that the order sets in abruptly below a critical temperature T_C has a close analogy with ferromagnetism', and going on to list 'the many points of similarity between the present treatment and the classical equation of Langevin and Weiss' (Bragg and Williams 1934: 707–8). All these mean-field theories found a place in the unified account of transition phenomena given by Landau in 1937.[19] It was referred to as the 'classical account' (not to distinguish it from a quantum-theoretic account, but to indicate its status).

Given these developments, how did the Ising model survive? Imre Lakatos, writing on the methodology of scientific research programmes, observes: 'One must treat budding programmes leniently' (Lakatos [1970] 1978: 92).

---

[16] Heisenberg 1928. The translation is from Stephen Brush 1967: 288. On the Heisenberg model, see Lee 1963: 114–15.

[17] van der Waals took the ideal gas laws, $PT = RT$, and modified the volume term to allow for the non-negligible volume of the molecules, and the pressure term to allow for mutual attraction. The resulting equation: $(P - a/V2)(V - b) = RT$, more nearly approximates the behaviour of real gases.

[18] On mean-field theories, in general, see Amit 1984: 6–8 and Domb 1996: 84–6. On anti-ferromagnetism, see Lee 1963: 202–4.

[19] Landau himself (1937) did not start from the analogy between pressure and magnetic field that I have described. Rather, part of his achievement was to show how that analogy emerged naturally from his analysis. See Domb 1996: 18; Pfeuty and Toulouse 1977: 25–8.

But, in 1935, the Ising approach to ferromagnetism was hardly a budding programme; it had withered on the vine. Any attempt to revive it, one might think, would have been a lurch from leniency into lunacy.

Yet, in 1967, Stephen Brush would write a 'History of the Lenz–Ising Model'; in 1981, a review article entitled 'Simple Ising Models Still Thrive' would appear, and its author, Michael Fisher, would announce two years later: 'one model which historically has been of particular importance ... deserves special mention: this is the Ising model. Even today its study continues to provide us with new insights' (Fisher 1983: 47). Nor was this view idiosyncratic. In 1992, a volume setting out recent results in the area of critical phenomena echoed Fisher's opinion: 'It is of considerable physical interest to study ... the nearest neighbor ferromagnetic Ising model in detail and with mathematical precision' (Fernandez, Fröhlich, and Sokal 1992: 6).

Ironically enough, a necessary first step in the resuscitation of the model was the recognition that Ising had made a mistake. While his proof that the linear, one-dimensional model does not exhibit spontaneous magnetization is perfectly sound, his conjecture that the result could be extended to models of higher dimensionality is not. In 1935, Rudolf Peierls argued that the two-dimensional Ising model exhibited spontaneous magnetization,[20] and, during the late 1930s, the behaviour of the model, now regarded as essentially a mathematical object, began to be studied seriously (see Brush 1967: 287). From its subsequent history, I will pick out just three episodes: first, the 'Onsager Revolution' of the 1940s (the phrase is Domb's); secondly, the emergence in the 1960s of the theory of critical exponents and universality; and, thirdly, the advent of renormalization theory in the 1970s.

6.1.5. The Onsager Revolution

Lars Onsager's achievement was to produce a rigorous mathematical account of the behaviour at all temperatures—including those in the immediate vicinity of T_C—of the two-dimensional Ising model in the absence of a magnetic field. His was a remarkable feat of discrete mathematics, an exact solution of a non-trivial many-body problem.[21] He showed, amongst other things

---

[20] Peierls 1936. The circumstances that occasioned Peierls' work on this topic were somewhat fortuitous.

[21] The essential problem was to calculate the partition function Z for the model. Recall that N is the number of sites in the lattice, assumed large. In any arrangement a there will be a certain number $N+$ of nearest-neighbour pairs $<j,k>$ for which $s_j(a) = +1, s_k(a) = -1$. The energy associated

(1) that the two-dimensional model exhibits spontaneous magnetization, but (2) that it does not occur above a critical temperature, T_C; further (3) that for the square lattice, $T_C = 2.269]/k$; and (4) that near T_C physical properties like the specific heat of the lattice show a striking variation with temperature. In his own words, 'The outcome [of this analysis] demonstrated once and for all that the results of several more primitive theories had features which could not possibly be preserved in an exact computation'.[22] I will come back to these results in the next section. Observe, however, that by 1970, when Onsager made these remarks, he could regard the mean-field theories of the classical approach as 'more primitive' than the mathematically secure Ising model. The whirligig of time had brought in his revenges. Concerning the immediate impact of Onsager's work, on the other hand, differing estimates are provided. Thus, whereas Domb declares that, '[The] result was a shattering blow to the classical theory', Fisher talks of 'Onsager's results which, in the later 1950s, could still be discarded as interesting peculiarities restricted to two-dimensional systems', and cite Landau and Lifshitz (1958: 438) in corroboration.[23] Landau, of course, was hardly a disinterested observer.

A complete and exact solution for the two-dimensional Ising model in an external field has not yet been achieved,[24] and an analytic solution of the three-dimensional Ising model is not possible. In both cases, however, very precise approximations can be made. With regard to the three-dimensional case, we may note that the approach pioneered by Domb, using the technique known as 'exact series expansion', involved comparatively early applications of computer methods to problems in physics.[25]

6.1.6. Critical Exponents and Universality

I have already mentioned some of the remarkable changes that occur in the physical properties of a system near its critical temperature. Particularly

---

with this arrangement of the model can be expressed as a function $g(N, N_+, N_{+-})$ (see Domb 1996: 112–15). In order to calculate Z, we need to calculate how many arrangements there are to each (N_+, N_{+-}). The solution of this combinatorial problem was Onsager's major achievement. A detailed account of this is given in Domb 1996: chap. 5.

[22] Lars Onsager, in autobiographical remarks made in 1970, quoted by Domb (1996: 130).

[23] Both quotations are from Domb 1996, the first from p.19 of the text, and the second from Fisher's foreword (xv).

[24] But, see Fernandez, Fröhlich, and Sokal 1992: 7–11 and chap. 14 for partial results.

[25] See Domb 1996: 22. For an outline of exact series expansion See Fisher 1983: 54–8.

significant are the variations with temperature of four quantities: the specific heat C, the order parameter M, the susceptibility χ, and the correlation length ξ. For expository purposes, it is often easier to adopt the idiom of ferromagnetism, to refer, for instance, to the order parameter M as the *spontaneous magnetization* (the magnetization in the absence of an external magnetic field). In this idiom, χ is the *magnetic susceptibility*, the ratio M/h of the magnetization to the applied magnetic field h. Of course, one could equally well interpret these quantities in terms of liquid-vapour critical phenomena, where the order parameter is the difference, $\rho - \rho_c$, between the actual density of the liquid-vapour fluid and its density at the critical point, and the susceptibility is the change of this difference with pressure (i.e. the fluid's compressibility). In the liquid-vapour case the relevant specific heat is the specific heat at constant volume (equivalently, at constant density, C_v).

For generality, we describe the variations of these quantities in terms of the *reduced temperature t*, defined as follows:

$$t = (T - T_C)/T_C \qquad (6.5).$$

On this scale, $t = -1$ at absolute zero, and $t = 0$ at the critical temperature. Clearly, the number of degrees Kelvin that corresponds to a one degree difference in the reduced temperature will depend on the critical temperature for the phenomenon under study.

Of the four quantities, the spontaneous magnetization decreases to zero as T_C is approached from below, and remains zero thereafter, but the other three all diverge (go to infinity) as T_C is approached from either direction (see Fig. 6.1). There is both experimental and theoretical support for the thesis that, close to T_C, the variation with t of each of these four quantities is governed by a power law; that is to say, for each quantity Q we have

$$Q \sim |t|^\lambda \qquad (6.6)$$

where, by definition, this expression means that

$$\lim_{t \to 0}[\ln(Q)/\ln(t)] = \lambda \qquad (6.7).$$

Given this definition, we see that, if $Q = Q_a|t|^\lambda$, where Q_a is some constant, then $Q \sim |t|^\lambda$. The converse, however, does not hold—a point whose importance will appear shortly.

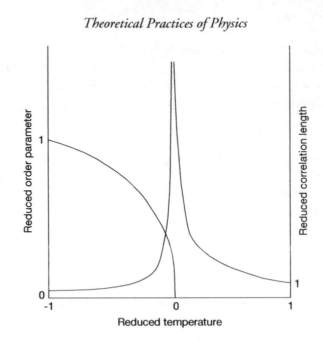

Fig. 6.1. The variation with temperature of the order parameter and correlation length of the two-dimensional Ising model. Reproduction from Paul Davies (ed.), *The New Physics*. Cambridge University Press, 1989: 241.

For the magnetization, M, we have, for $t < 0$,

$$M \sim |t|^\beta \tag{6.8}$$

and, for the three diverging quantities,

$$C \sim |t|^{-\alpha} \qquad \chi \sim |t|^{-\gamma} \qquad \xi \sim |t|^{-\upsilon}$$

The exponents β, α, γ, and υ are known as *critical exponents*.

The behaviour of the last three quantities is not, as these relations might suggest, symmetrical about T_C where $t = 0$ (see Fig. 6.1). In fact, at one time it was thought that different exponents were needed on either side of T_C, so that we should write, for example,

$$\xi \sim |t|^{-\upsilon} \text{ for } t > 0 \qquad \xi \sim |t|^{-\upsilon'} \text{ for } t < 0 \tag{6.9.}$$

Table 6.1. **Exponents for the two-dimensional Ising model**

Exponent	Ising Model ($d = 2$)	Ising model ($d = 3$)	Mean-field theory
α	0 (log)	0.12	0
β	1/8	0.31	1/2
γ	7/4	1.25	1
υ	1	0.64	1/2
δ	15	5.0	3
η	1/4	0.04	0

Source: Data from Amit 1984: 7. Amit (p. 235) gives the current estimate for β and δ in the three-dimensional Ising model as 0.325 and 4.82, respectively. Unfortunately, he does not supply a comprehensive list.

There are now, however, strong reasons to think that $\alpha = \alpha'$, $\gamma = \gamma'$, $\upsilon = \upsilon'$ and to look elsewhere for the source of asymmetries. In a simple case we might have, for instance,

$$Q \sim Q_+ |t|^\lambda \qquad Q \sim Q_- |t|^\lambda \qquad (6.10)$$

where $Q_+ \neq Q_-$, but the exponents are the same.

Two other critical exponents, δ and η, are defined in terms of the variation *at the critical temperature* of M with h, and of the correlation function, $f(R)$ with R. We have,

$$M \sim h^{1/\delta} \qquad \Gamma(R) \sim R^{-(d-2+\eta)} \qquad (6.11).$$

In the latter, d denotes the dimension of the physical system.

In Table 6.1, the exponents have been calculated exactly for the two-dimensional Ising model.[26] These values are compared with the approximate values obtainable for the three-dimensional version, and the values according to mean-field (Landau) theory.

The fact that $\alpha = 0$ for both the two-dimensional Ising model and the mean-field theory masks a radical difference between the two. In 1944,

[26] In saying this I am assuming the scaling laws which were used to establish the exact value δ (see Table 6.1); however, precise estimates of the approximate value for δ were available consistent with the exact value shown. See Domb 1996: 175.

Onsager proved that in the two-dimensional Ising model, C would diverge logarithmically as the temperature approached T_C from either direction; from above ($t > 0$) we have:

$$C = A \ln(t) + c \tag{6.12}$$

where c is a finite 'background' term. When the formal definition (6.7) of the $\sim$-relation is applied, this yields:

$$\alpha = \lim_{t \to 0} \ln[\ln(C)]/\ln(t) = 0. \tag{6.13}$$

Hence the use of '0(log)' in the table. In contrast, on the mean-field theory, the specific heat remains finite. It rises almost linearly whether T_C is approached from above or below. The zero exponent appears because there is a discontinuity at T_C. Denoting by $C^+(C^-)$ the limiting value of C we have $C^- > C^+$ approached from above or from below.

Note also that there is no other exponent for which the mean-field value coincides with the value for either Ising model. These differences are much more threatening to the mean-field theory than might be supposed. One might think that the situation was symmetrical: given two competing models, experiment would decide between them. But, in this case, the two theories are different in kind. Mean-field theory, as envisioned by Landau, was a universal theory of critical phenomena. He used symmetry considerations and well-established thermodynamic principles to generate results which he hoped would be applicable to any system exhibiting critical-point behaviour. Now, the two-dimensional Ising model may not be an entirely faithful representation of any physical system with which we are familiar. None the less, any system that it *did* represent faithfully would exhibit critical phenomena, and so come under the umbrella of mean-field theory. Thanks to Onsager, we now know precisely what the behaviour of such a system would be, and that it would not conform to the mean-field theory's predictions.[27]

---

[27] The problem with the Landau theory can, in retrospect, be described in two ways. To put it formally, Landau expressed the free energy of a system in terms of a power series of M^2: the square is used for reasons of symmetry; he then ignored terms above the M^2 and M^4 terms. However, at T_c the series does not converge. To put this physically, what Landau ignored were the fluctuations: regions of varying degrees of order that occur at T_c. See Amit 1984: 7.

The 1960s saw the emergence of a new general approach to critical phenomena, with the postulation of the so-called *scaling laws*, algebraic relationships holding between the critical exponents for a given system. Five of them are given by equations (a)– (e) below. Listed alongside them are the more fundamental inequalities of which the equations are special cases.

(a) $\alpha + 2\beta + \gamma = 2$	(a*) $\alpha + 2\beta + \gamma \geqslant 2$	(Rushbrooke)	(6.14)
(b) $\alpha + \beta(1 + \delta) = 2$	(b*) $\alpha + \beta(1 + \delta) \geqslant 2$	(Griffiths)	(6.15)
(c) $v(2 - \eta) = \gamma$	(c*) $v(2 - \eta) \geqslant \gamma$	(Fisher)	(6.16)
(d) $dv = 2 - a$	(d*) $dv \geqslant 2 - a$	(Josephson)	(6.17)
(e) $d[(\delta - 1)/(\delta + 1)]$ $= 2 - \eta$	(e*) $d[(\delta - 1)/(\delta + 1)]$ $\geqslant 2 - \eta$	(Buckingham and Gunton)	(6.18)

In equations (d) and (e), and in the corresponding inequalities, d is the dimensionality of the system. The equations (a)–(e) are not independent; given (a)–(c), (d) and (e) are equivalent. The inequalities, however, are algebraically independent, and, indeed, were independently proposed by the individuals cited next to them. These inequalities do not all have the same warrant. The first two listed are derivable from the laws of thermodynamics, specifically from the general thermodynamic properties of the free energy of a system, while the others depend on additional assumptions about correlation functions $\Gamma(r)$, which hold for a very broad class of systems.[28] The additional postulates needed to yield the individual equations are known as the *thermodynamic scaling hypothesis* (for (a) and (b)), the *correlation scaling hypothesis* (for (c)), and the *hyperscaling hypothesis* (for (d) and (e)).[29]

All five scaling laws hold exactly for the two-dimensional Ising model, provided that the exact value 15 is assigned to δ (see n. 26). They also hold approximately for the three-dimensional Ising model (the largest error, given the Amit values, is 3 per cent), but that is not surprising, since conformity with these relations was one of the criteria by which the approximations used in calculating the exponents were assessed (see Domb 1996: 175). For the mean-field theory, equations (a), (b), and (c) all hold, but neither the equations (d) and (e) nor the inequalities (d*) and (e*) hold, unless $d = 4$.

---

[28] For an extended discussion of the inequalities, see Stanley 1971: chap. 4.

[29] A useful taxonomy of the various equations and inequalities is given by Fernandes, Fröhlich, and Sokal 1992: 51–2. For a full discussion of the scaling hypotheses, see Fisher 1983: 21–46.

Thus, in 1970, the situation was this. The differences between the values of the critical exponents calculated for the two-dimensional Ising model, the three-dimensional Ising model, and the mean-field model had suggested not only that the mean-field theory was not a universal theory of critical phenomena but also that the search for one was perhaps mistaken. But then, with the postulation of the scaling laws, a new twist had been given to the issue of universality. Unlike the mean-field theory, the laws did not suggest that the critical exponents for different systems would all take the same values; still less did they prescribe what those values might be. What they asserted was that the same functional relations among the critical exponents would obtain, no matter what system was investigated, provided that it exhibited critical-point behaviour. More precisely, given a system of a certain dimensionality, if we knew two of these exponents, β and v, then from the laws we could deduce the other four.

Yet, intriguing though these laws were, in 1970, the status of the various scaling hypotheses, and hence of the laws themselves, was moot. The thermodynamic scaling hypothesis, for instance, could be shown to follow from the requirement that the Gibbs potential for a system exhibiting critical-point behaviour be a generalized homogeneous function (for details, see Stanley 1971: 176–85). This requirement, however, is entirely formal; it has no obvious physical motivation. As Stanley commented (ibid. 18): '[T]he scaling hypothesis is at best unproved, and indeed to some workers represents an *ad hoc* assumption, entirely devoid of physical content.' Nor did physically based justifications fare better. Introducing a rationale for adopting the hyperscaling hypothesis, Fisher cheerfully remarks (1983: 41–2): 'The argument may, perhaps, be regarded as not very plausible, but it does lead to the correct result, and other arguments are not much more convincing!'

Experimental verification of the laws was difficult since the task of establishing precise values for critical exponents was beset by problems. Two examples. First, in no finite system can the specific heat or the correlation length increase without bound; hence their variation with the reduced temperature very close to T_C will not obey a power law; and the relevant critical exponent will not be defined (see Fisher 1983: 140). Secondly, the reduced temperature t that appears in relation (6.8): $M \sim |t|^{\beta}$ is given by $t = (T - T_C)/T_C$. But, T_C may not be known in advance. Hence, if we use a log-log plot of the data to obtain β from $\beta = \lim_{t \to 0}(\ln M)/(\ln - t) = d(\ln M/)d(\ln - t)$, a series of trial values of β may have to be assumed until a straight line is produced.

That is to say, prior to obtaining the exponent β we will need to *assume* that a power law governs the relation between M and t (see Stanley 1971: 11).

Again, if the scaling laws are to be justified by appeal to known critical exponents, then those exponents had better not be established via the scaling laws. Yet, Domb notes (1996: 174), concerning the calculations performed on various three-dimensional lattices in the 1960s: 'It was usual to make use of the [Rushbrooke] scaling relation ... and the well-determined exponents γ and β to establish the value of α', and Fisher writes (1983: 41), perhaps with tongue in cheek: 'Experimentally also [the Fisher equality] checks very well. If it is accepted, it actually provides the best method of measuring the elusive exponent η!'[30]

Throughout the 1970s and 80s, however, theoretical and experimental evidence pointed in support of the scaling laws. Furthermore, this evidence pointed to an unanticipated conclusion: that systems exhibiting critical behaviour all fall into distinct classes, within which the values of the critical exponents of every system are the same.[31] These 'universality classes' are distinguished one from the other by two properties: the dimension of the systems involved and the symmetry of its site variable. In the case of the Ising model, for example, where $s_j = \pm 1$, the symmetry is that of the line, which is invariant under reflection; in the case of the Heisenberg model of isotropic ferromagnetism, on the other hand, the site variable is a three-component spin, and thus has spherical symmetry.[32] The study of critical-point phenomena had revealed a remarkable state of affairs. While critical temperatures are specific to individual phenomena, as are the amplitudes of quantities like the order parameter and the correlation length, the power laws that govern the variation of these quantities with temperature are determined by dimension and symmetry, properties that radically different systems can have in common.

The study of universality within critical-point phenomena gave the Ising model a role different from any it had previously played. In 1967, when Brush wrote his 'History of the Lenz–Ising Model', the model's chief virtues were seen as its simplicity and its versatility. As Brush notes, and we have seen in section 6.1.3, its versatility was evident from the direct, though

[30] If I have lingered unduly on the topic of the scaling laws and their justification, it is because they would provide wonderful material for a detailed case study of justification and confirmation in twentieth-century physics.

[31] For theoretical evidence, see, e.g. Griffiths 1970 and Domb 1996: chap. 6. For experimental evidence, see Balzarini and Ohra 1972 on fluids, discussed by Fisher (1983: 8), and, especially, the review article by Ahlers (1980). [32] For other examples, see Amit 1984: 8–9.

crude, representations it offered of the atomic arrangements within systems as diverse as ferromagnets, liquid-vapour mixtures, and binary alloys.

Another virtue was its mathematical tractability. At least for the two-dimensional version, exact values for critical exponents were calculable; thus, the model could fulfil the negative function of providing a counter-example to the mean-field theory's predictions, and the positive one of confirming the scaling laws. And, a propos of results obtained for the three-dimensional Ising model, Domb also suggests, perhaps with hindsight, that, 'The idea that *critical exponents for a given model depend on dimension and not on lattice structure* was a first step towards the *universality hypothesis*' (1996: 171; emphasis in the original).

Be that as it may, with the enunciation and acceptance of that hypothesis came a new emphasis on the use of models in condensed matter physics. By definition, universal behaviour supervenes on many different kinds of processes at the atomic or molecular levels. Hence, its study does not demand a model that faithfully represents one of those processes. To quote Bruce and Wallace:

The phenomenon of universality makes it plain that such details are largely irrelevant to critical-point behaviour. Thus we may set the tasks in hand in the context of simple model systems, with the confident expectation that the answers which emerge will have not merely qualitative but also quantitative relevance to nature's own systems. (Bruce and Wallace 1989: 242; emphasis in the original)

In other words, in this field a good model acts as an exemplar of a universality class, rather than as a faithful representation of any one of its members.

The virtues we have already remarked enable the Ising model to fit the role perfectly. Its mathematical tractability and—even more—its amenability to computer simulation allow it to be used to explore critical-point behaviour with great precision and in great detail. In addition, however crude its representation of ferromagnetism may be at the atomic level, the fact that it was devised with a particular phenomenon in mind provides an added benefit. It enables physicists to use the physical vocabulary of ferromagnetism, to talk of spins being aligned, for example, rather than of site variables having the same value, and it encourages the play of physical intuition, and in this way it facilitates understanding of critical-point phenomena.[33] It is not surprising that Kenneth Wilson's 1973 lectures on the renormalization

---

[33] On models and understanding, see Fisher 1983: 47; for an account of explanation in terms of models, read Essay 7.

group include a lecture entirely devoted to the Ising model (Wilson 1975: 797–805).

My discussion of the renormalization group in the next section is nearly all in terms of the Ising model. It thus offers *inter alia* an illustration of how heuristically useful the model can be.

6.1.7. Universality and the Renormalization Group

To understand universality, one must remember two things: first, the fact that the term is applied to the behaviour of systems at or near the critical temperature; secondly, that the correlation length ξ has a particular significance at that temperature.

Of the six critical exponents discussed in the previous section, α, β, γ, and υ appear in power laws that describe the behaviour of physical quantities at or near the critical temperature, while δ and η appear in laws describing behaviour at or near the critical temperature itself. Amongst the former is the relation $\xi \sim |t|^{-\upsilon}$. It tells us that the correlation length ξ tends to infinity as the system approaches T_C. As I noted in section 6.1.2, in the Ising model the correlation length can be thought of as a measure of the maximum size of a totally ordered island within the lattice as a whole. Not all totally ordered regions, however, are of this size; in fact, at the critical temperature, islands of all sizes will be present, ranging from those whose characteristic length is equal to ξ all the way down to those containing a single site—that is, those whose characteristic length is given by the lattice spacing.

This prompts the hypothesis that universality results from the fact that islands of all these different sizes are involved in the physics of critical-point behaviour. Put in terms of the Ising model, it states that between the lattice spacing and the correlation length there is no privileged length scale. It thus gives rise to two lines of investigation. The first is straightforward: the investigation of what resources the Ising model possesses for providing a description of the lattice in terms of regions rather than individual sites. The second is conceptually subtler: an enquiry into how state variables associated with these larger regions could reproduce the effects of the site variables s_i. At the site level, the tendency for the site variables of adjacent sites to become equal (or, in magnetic terms, for their spins to become aligned) is described in terms of interaction energies between nearest neighbours. By analogy, we may enquire whether there is a description available at the regional level that describes the coupling strengths between one region and another. Whereas we normally think of the nearest-neighbour interactions as ultimately responsible for correlations in the lattice at all ranges from the lattice spacing up to the correlation length, this approach suggests that the coupling strengths between regions of any characteristic length L may be regarded as fundamental, in the sense that they give rise to correlations at all ranges between L and ξ.[34]

Both approaches shed light on the problem of universality. To illustrate the first, consider the example of the Ising model defined in section 6.1.2, which uses a two-dimensional square lattice of sites. This lattice can equivalently be regarded as an array of square cells, or *blocks*, with sides equal in length to the lattice spacing. Now, suppose that we obtain a coarse-grained picture by doubling the length of the sides of the blocks. Each block will then contain four lattice sites. Clearly, the new array of blocks itself constitutes a lattice, and we may define block variables on it by adding the site variables from the sites it contains. These block variables will take one of five values: -4, -2, 0, $+2$, $+4$. The coarse-graining procedure can be repeated, each step doubling the lengths of the blocks' sides. After n such dilations each block will contain a $2^n \times 2^n$ array of sites. (We can regard the formation of the blocks that contain a single site as step zero.) After n steps, the block variable will take one of $(2^n)^2 + 1$ values.[35]

We may scale these block variables to run from -1 to $+1$; as n increases they tend towards a continuum of values between these limits. If the values of the site variables were independent of one another, as happens at high temperatures, where they are randomized by thermal noise, we would expect the probabilities of these variables over the lattice to form a Gaussian distribution centred on zero. At the critical temperature, however, the effect of correlations is that, as n increases, the probability distribution of the block variables, or *configuration spectrum*, tends towards a bimodal form with peaks at roughly ± 0.85, as shown in Fig. 6.2. The value of n for which the probabilities for a discrete set of values fit this curve is surprisingly small. Bruce and Wallace show (1989: 247) how good the fit is for $n = 3$, when there are sixty-five possible values of the block variable, and thereby show that for all lengths greater than eight lattice spacings the probability distribution of (scaled) block variables is a *scale-invariant property*. They also use the same

[34] The two approaches correspond to Bruce and Wallace's *configurational view* and *effective coupling view* of the renormalization group, and I have abridged their treatment in what follows.
[35] It is not necessary that we define block variables in this way. Niemeijer and van Leeuwen (1974) consider a triangular lattice.

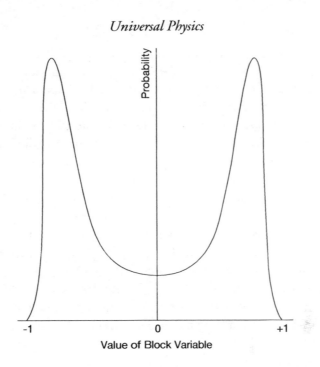

Fig. 6.2. Configuration spectrum for the two-dimensional Ising model.

plot to show an even more striking fact: that this configuration spectrum is independent of the number of values of the site variables in the original model. They do so by taking an Ising model whose site variable can take one of three different values $(-1, 0, +1)$ and plotting the probabilities of the 128 different block values obtained when $n = 3$. Whereas the local configuration spectra of the 2-value and the 3-value models are quite different, these differences get washed out, as it were, under coarse-graining.

Taken together, the two results corroborate the link between universality and scale invariance. Wallace and Bruce extend this analysis by an ingenious argument to show that the way in which the configuration spectrum evolves under coarse-graining is characterized by the critical exponents β and υ. This result is entirely consistent with the conclusions quoted at the end of the last section (1) that the values of β and υ suffice to determine the values of all critical exponents for a given universality class, and (2) that the physical properties that define a particular class are the dimensionality of the system and the symmetry of its state variable. Both of the models in question are two-dimensional, and the change from a two-valued to a three-valued state variable is not a change of symmetry, since in each case the symmetry is linear.

The results about configuration spectra that I have quoted were taken from computer simulations of the behaviour of the Ising model. Given the experimentally confirmed fact of universality, we may expect them to hold for any two-dimensional system whose state variable has linear symmetry. But, although these simulations show that the Ising model does indeed display scale-invariant properties at its critical temperature, the question remains how this scale-invariance comes about. To answer this question, the second of the two approaches I suggested earlier is called for.

On this approach, any level of description for which the characteristic length of the resolution lies between the lattice spacing and the correlation length could be regarded as fundamental. That is to say, the macroscopic behaviour of the model could be thought of either as produced by nearest-neighbour interactions between sites in the standard way, or as generated by interactions between blocks, provided that the length of the blocks' edges was small compared with macroscopic lengths. Two questions arise. First, what kinds of interaction between blocks would produce the same results as the nearest-neighbour interactions between sites? And, secondly, what sort of relation between one set of interactions and the other could be generalized to blocks of different edge-lengths?

Before addressing these questions, let me note that, in the Ising model, the effects of the interactions between sites are attenuated by temperature. In the key equation (6.3) that governs the statistical behaviour of the lattice, probabilities depend, not just on the interaction energy E_a of an arrangement, but on E_a/kT. This attenuation can be absorbed into the description of the interactions between sites by specifying them in terms of a coupling strength K, where

$$K^0{}_1 = J/kT \qquad (6.19).$$

Here the superscript '0' indicates that we are regarding a site as a block produced at the zero step in the coarse-graining procedure, and the subscript '1' registers the fact that the interactions are between nearest neighbours. The point of the second notational device is this. When we ask what effective interactions exist between blocks produced after n steps of coarse-graining, there is no guarantee that they will be confined to nearest-neighbour inter-actions. In fact, they may have to be represented by a set of effective coupling strengths, $K^n{}_1, K^n{}_2, K^n{}_3$, etc., where $K^n{}_1$ represents a coupling strength for adjacent blocks, $K^n{}_2$ a coupling strength between blocks two block-lengths apart, and so on. The superscript n indicates that the blocks have been produced after n steps of the coarse-graining procedure, so that the blocks' sides are of length 2^n lattice spacings. A more economical representation is obtained if we think of a coupling strength as a vector $\mathbf{K}^n$ whose components are $K^n_1, K^n_2, \ldots.$ The vector $\mathbf{K}^0$ has only one component, K^0_1.

From the coupling strength $\mathbf{K}^0$, n steps of coarse-graining yields the effective coupling strength $\mathbf{K}^n$. We may represent the effect of this coarse-graining mathematically by an operator T^n on the space of coupling strengths,[36] such that

$$\mathbf{K}^n = T^n(\mathbf{K}^0) \tag{6.20}.$$

Various physical assumptions have already been built into this mathematical representation. It is assumed that the model is describable in terms of the states of blocks of arbitrary size; furthermore, that whatever level of coarse-graining we choose, the internal energy of the model is expressible in terms of a set of effective coupling strengths associated with interactions between blocks at different separations, these separations always being specified in terms of the block spacing. In brief, built into the representation is the hypothesis we encountered earlier, that between the lattice spacing and the correlation length there is no privileged length scale.

The further requirement, that the interactions between regions of any characteristic length L may be regarded as fundamental—that is, as generating the correlations at all lengths between L and ξ, constrains the family $\{T^n\}$ of transformations. We require, for all m, n:

$$T^{n+m} = T^m.T^n \tag{6.21}.$$

This equation is satisfied if, for all n:

$$\mathbf{K}^{n+1} = T^1(\mathbf{K}^n) \tag{6.22}.$$

Equation 6.21 is the rule of group multiplication for an additive group. Hence, the name, 'Renormalization Group', for $\{T^n\}$.[37] It is a sequence of transformations induced by a sequence of dilations. Like many groups of

---

[36] I distinguish the operator T^n from the temperature T by italicizing the latter T.

[37] Strictly, $\{T^n\}$ for a semigroup; no inverse operation is employed.

transformations in physics, the transformations induced by coarse-graining can leave the system looking different, particularly when n is small, while its fundamental physical properties remain undisturbed.[38]

The possibility of universal behaviour appears if the sequence: K^0, K^1, K^2, K^3, … tends to a limit K^*. In this case, the transformation T^1 has a *fixed point* K^* such that

$$T^1(K^*) = K^* \tag{6.23}.$$

The fixed point is not determined by a unique sequence: K^0, K^1, K^2, … but by the transformation T^1. In other words, there are many sequences that tend to the same fixed point; in physical terms, there may be many systems which differ at the atomic level, but which nevertheless manifest the same behaviour at large length scales. Furthermore, when n is large the points K^n, K^{n+1}, K^{n+2}, … lie very close together, and are also very close to K^*; in physical terms, systems display scale-invariant behaviour at large length scales.

This skeletal account of renormalization theory has done no more than show how universality might come about. The theory itself changed the way physicists approach critical-point phenomena (see Domb 1996: chap. 7). Its articulation by Kenneth Wilson in the early 1970s earned him a Nobel prize a decade later. It is an unusual theory. It is not mathematically rigorous, and the approximative methods used are chosen on a case-by-case basis. That is part of the reason why simple mathematical models are so valuable in foundational work. In Wilson's own words, for physicists seeking to apply the theory, 'There is no renormalization cook book'.[39]

6.1.8. Computer Simulation of the Ising Model

The use of computers to perform calculations has been commonplace in critical-point physics for fifty years. I have already mentioned (section 6.1.5) their use by Domb and his co-workers in establishing the behaviour of the three-dimensional Ising model. Some fifteen years later, in applying the renormalization approach to the two-dimensional Ising model, Wilson also used the computational capacity of the computer. Indeed, in describing what

---

[38] Naturally, what counts as fundamental depends on one's preferred group of transformations.
[39] Wilson, quoted in Domb 1996: 261.

he achieved, he tells us about the computer he used (a CDC 7600) and the details of the calculations involved (Wilson 1975: 802–5).

Yet, although Domb and Wilson both used computers, and although they both worked on the Ising model, to my mind neither of them performed a computer simulation of the model's behaviour.[40] A typical computer simulation of the two-dimensional Ising model involves the representation of an array of sites within the computer in such a way that values of the site variable can be attached to each site. A stochastic algorithm takes one arrangement a of the lattice into another, a', with a specified probability. Repeated applications of the algorithm generate a sequence $a, a', a'', \ldots$ of arrangements. The algorithm is so designed that, during a suitably long sequence, arrangements will occur with a relative frequency equal to their probability according to the Boltzmann Law:

$$p_a = Z^{-1} \exp(-E_a/kT) \tag{6.3}.$$

In brief, under the algorithm the set of arrangements becomes a canonical ensemble. Properties of the model are measured by taking mean values (as determined by the computer) over a suitably large number of arrangements.

This type of simulation is called a Monte Carlo simulation because, like a roulette wheel, it chooses outcomes from a set of alternatives with specified probabilities. The simulation of the three-dimensional Ising model devised by Pearson, Richardson, and Toussaint (1983) is a nice example.[41] In the basic step of their algorithm, a particular site i is considered, along with the values $s_n(a)$ of the site variables of six nearest neighbours. The value of the variable s_i is then set to $+1$ with probability

$$p = \exp(-E_n/kT)/[\exp(-E_n/kT) + \exp(E_n/kT)] \tag{6.24}$$

and to -1 with probability $1 - p$.[42] (Here E_n is an abbreviation for $-J\Sigma s_n(a)$, where the summation is over the six neighbouring sites s_n of i.[43]) This step

---

[40] I treat this distinction at greater length in section 6.2.2.

[41] For an introduction to Monte Carlo methods, and a discussion of their applications to critical-point phenomena, see Blinder and Stauffer 1984 and Blinder 1984a, respectively. Both contain sections on the simulation of the Ising model; neither paper, however, does more than mention single-purpose processes of the kind that Pearson, Richardson, and Toussaint constructed.

[42] The Monte Carlo process uses a random generator to produce a number x in the interval [0,1]. This number is then compared with p. If $x<p$ the value $+1$ is selected; otherwise the value -1 is selected. See Blinder and Stauffer 1984: 6–7.

[43] For generality, Pearson, Richardson, and Toussaint include in this energy term an additional h to allow for an external field.

is then reiterated for each site in the lattice in turn, and the sweep over all sites is repeated many times. It can be shown that, whatever the initial arrangement, the simulation will approach an equilibrium situation in which the Boltzmann probabilities are realized.

Pearson and his collaborators modelled a cubic lattice with $64 \times 64 \times 64$ sites. With $1/kT$ set at 0.2212 ($1/kT_C = 0.2217$ for this lattice), the equilibrium situation is reached after about 1,700 sweeps through the lattice. As they point out:

This means that the great bulk of the computational effort consists of repeating the simple updating algorithm for all the spins in the lattice. Therefore it is attractive to construct a special purpose device to perform the updatings and make some of the simplest measurements.   (Pearson, Richardson, and Toussaint 1983: 242)

With this device they performed the Monte Carlo updating on 25 million sites per second; thus each second roughly 100 sweeps of the lattice were performed. 'Despite its modest cost', they claim (ibid. 241), 'this machine is faster than the fastest supercomputer on the one particular problem for which it was designed'.[44]

It was possible to adjust the machine so that the coupling strength in the x-direction was set to zero. It then behaved like an array of two-dimensional lattices, all of which were updated in one sweep of the model. In this way the verisimilitude of the simulation could be checked by comparing the performance of the machine against the exactly known behaviour of the Ising model. When their respective values for the internal energy of the lattice at various temperatures were compared, they agreed to within the estimated errors for the values given by the simulation; that is, to within 1 part in 10,000.[45]

The main function of the processor, however, is to simulate the behaviour of the three-dimensional Ising model, for which exact results are not available. For this reason, Pearson et al. display plots of the variation with temperature of the magnetization, susceptibility, and specific heat of the model as examples of what the simulation can provide.

Computer simulations, in the strict sense, of the behaviour of the Ising model started to be performed in the 1960s. In his well-known text, Stanley

---

[44] In assessing this claim, one should bear in mind when it was made. Note also, that while a third machine was being built in San Diego, California another single-purpose machine using the Ising model was being built in Delft by Hoogland and colleagues. See Hoogland et al. 1983.
[45] See Pearson, Richardson, and Toussaint 1983: 247–8.

(1971: facing p. 6) illustrates his discussion of the lattice-gas model with some pictures generated by a simulation performed by Ogita and others. Six pictures are displayed, each showing a 64×64 array of square cells, some black, some white. Stanley's remarks in the caption are a trifle odd, given that the Ising model can be used to model the lattice gas:

Fig. 1.5. Schematic indication of the lattice-gas model of a fluid system. [*There now follow notes about the individual pictures.*] This illustration and the associated temperatures are to be regarded as purely *schematic*. In fact the figure was constructed from a computer simulation of the time-dependent aspects of the two-dimensional Ising model and actually represents rather different phenomena.　(ibid.)

The distinction drawn here, between a *schematic indication* and a *representation* is a nice one. But, it may be that Stanley's chief concern is the difference of dimensionality between the two models.

Obviously, for expository purposes simulations of the two-dimensional version of the Ising model are ideal, since they can be presented on a two-dimensional grid or, in other words, as pictures. Yet, as Bruce and Wallace point out (1989: 239 n.), pictures like these are in one way misleading. A picture will show just one arrangement, but the properties that concern us are averaged over a sequence of arrangements. It follows that a 'typical' picture will have to be selected for the viewer.

Lending itself to pictorial display is not a prerequisite for a simulation. None the less, it helps. The pictures that illustrate Bruce and Wallace's presentation of renormalization theory allow many features of critical behaviour to be instantly apprehended, like the almost total disorder above the critical temperature, the high degree of order below it, and, crucially, the presence of ordered islands of all sizes at the critical temperature itself. It also allows the authors to display the results of coarse-graining at these different temperatures, to show that coarse-graining appears to take a model that is not at the critical temperature further away from criticality, so that a model above the critical temperature appears more disordered, and one below it more strongly ordered. At the critical temperature, on the other hand, provided the coarse-grained pictures are suitably scaled, the model looks no more and no less ordered after coarse-graining than before.

Rumination on the different ways in which changes in the pictures occur brings out a perplexing aspect of the renormalization process. We may distinguish between changes in the pictures that result from altering one of the parameters that govern the algorithm (in particular, the temperature), and changes that result from taking an existing picture and transforming it (in particular, by coarse-graining). When we move from study of the Ising model to the study of physical systems, this raises a puzzling question for the realist: How is she or he to regard the effective coupling strengths that are invoked when the length scale at which we describe the system is increased? On the one hand, they must surely be regarded as real; large regions of the system do have an influence one on the other. On the other hand, they seem like artefacts of the mode of description we choose to employ. A third alternative, which would have delighted Niels Bohr, is that, while there can be no single complete description of a system near to its critical point, renormalization theory shows how a set of complementary descriptions can be obtained. I will not pursue this issue here.

6.2. THEORETICAL REPRESENTATION: THE ISING MODEL AND COMPUTER SIMULATION

6.2.1. The DDI Account of Modelling

The Ising model is employed in a variety of ways in the study of critical-point phenomena. To recapitulate, Ising proposed it (and immediately rejected it) as a model of ferromagnetism; subsequently it has been used to model, for example, liquid-vapour transitions and the behaviour of binary alloys. Each of these interpretations of the model is in terms of a specific example of critical-point behaviour. But, as we saw in the discussion of critical exponents, the model also casts light on critical-point behaviour in general. Likewise, the pictures generated by computer simulation of the model's behaviour illustrate not only ferromagnetic behaviour close to the Curie temperature but also the whole field of scale-invariant properties. In this section of the essay and the next I will see how the multiple roles played by the model and its simulation can be accommodated within my very general account of theoretical representation given in Essay 5, and which I called the *DDI account*.[46]

The account rests on the premiss that the function of a model is to provide a representation of a phenomenon, or a cluster of phenomena. The question is how best to characterize the kind of representation that a theoretical model

---

[46] I discuss this account at greater length in Essay 5.

provides. The DDI account invites us to think of a theoretical representation in terms of three components: denotation, demonstration, and interpretation (see Fig. 5.3). Elements of the subject of the model (a physical system evincing a particular kind of behaviour, like ferromagnetism) are *denoted* by elements of the model; the internal dynamic of the model then allows conclusions (answers to specific questions) to be *demonstrated* within the model; these conclusions can then be *interpreted* in terms of the subject of the model.

The DDI account thus shares with Goodman's account of pictorial representation the negative thesis that (*pace* Giere) representation does not involve a similarity or resemblance between the representation and its subject, and the positive thesis that elements of the subject are denoted by elements of the representation (see Goodman 1968: chap. 1). On the DDI account, however, theoretical representation is distinguished from pictorial and cartographic representation by the second component in the process, the fact that models, at least as they are used in physics, have an internal dynamic which allows us to draw theoretical conclusions from them. We talk about the behaviour of a model, but rarely about the behaviour of a picture or a map. The conclusions we draw are, strictly speaking, about the behaviour of the model; they need interpretation if they are to be applied to the phenomena under study.

The two-dimensional Ising model, for example, is an abstract mathematical entity. Hence, to paraphrase Goodman (1968: 5), it resembles other mathematical entities much more than it resembles, for example, a ferromagnetic layer on the surface of a physical object. When the model is used as a model for that kind of physical system, the representation it provides is generated by specific denotations. The values of s_i in the model denote the orientations of the elementary magnets that are assumed to make up the layer. Likewise, the parameter T in the model denotes the temperature, not of the model, but of the layer being modelled. As T increases, the model does not become hot to the touch.

Notice that, in talking, as I did, of 'the elementary magnets that are assumed to make up the layer', I am already using what Nancy Cartwright has called a 'prepared description' of the model's subject (1983: 133–4), and the use of a prepared description is even more obvious when we employ the (three-dimensional) Ising model as a model of a fluid (see (b) in section 6.1.1 and 6.1.3 above). These prepared descriptions are themselves representations, and so in these cases the Ising model is a representation of a representation. In fact, representational hierarchies of this kind are the norm within physics,[47] and, as we shall see later, they can easily be accommodated within the DDI account.

More problematic for the DDI account than these individual cases is the use of the Ising model in discussions of critical-point phenomena in general. The universality thesis suggests that Onsager's analysis of the two-dimensional Ising model has considerable theoretical significance, whether or not the model is a faithful representation of any specific example of critical-point behaviour—that is, whether or not the elements of the model denote elements of any actual physical system. In some discussions (e.g. Fisher 1983), the Ising model comes to represent a large, and very disparate class of physical systems almost in the manner that a congressman represents a district, or member of parliament a constituency. It represents a class of systems by being a representative of them. (It is also representative of them, which is not quite the same thing.) In such cases of representation, the concept of denotation seems out of place.[48]

Three possible responses to this problem suggest themselves. The first is to dismiss it, to say that the Ising model occupies a unique position in physics, and that since no other model acts as a representative of such a heterogeneous class of systems, no general account of modelling can be expected to accommodate it. The second response is less cavalier. It follows the line I took earlier in assessing the importance for mean-field theory of Onsager's results. In that case, I argued, since the mean-field theory's claims were purportedly universal, it did not matter whether the two-dimensional Ising model faithfully represented any actual physical system. It was enough that we could describe a physically possible fictional system that the model *could* represent, a fictional system, in other words, whose elements *could* be denoted by elements of the model.

The third response is, I think, more interesting. It requires us to isolate those features or characteristics that physical systems in a given universality class have in common, and which give rise to the scale-invariant behaviour of the members of that class. While these characteristics are not confined to the dimensionality of the systems involved and the symmetry of the physical quantities that govern their internal energy, neither do they include specific details of microprocesses, since these are not shared by all members of the

---

[47] Likewise, these hierarchies are discussed in some detail in Essay 5.
[48] Goodman explicitly excludes these cases from his discussion (1968: 4).

class. To each of them, however, there must correspond a characteristic of the Ising model, if it is to act as a representative of the class. Because of the generality of the predicates involved (e.g. 'forms a canonical ensemble'), we may be reluctant to use the term 'denotation' for the relation of correspondence between the characteristics of the model and of the individual physical systems. None the less, we may say that the existence of these correspondences allows us to regard the model as a whole as denoting each system as a whole, without thereby stretching the notion of denotation beyond recognition.

In such a case, the term *interpretation* needs to be similarly generalized. Not every detail of the Ising model can be directly interpreted as denoting a detail of a system in the class. In other words, the Ising model is not a faithful model of any such system. Yet it may well be that the behaviour of the model offers the best available explanation of the behaviour of one of the systems. As Fisher writes, immediately before he introduces the Ising model:

[T]he task of the theorist is to understand what is going on and to elucidate which are the crucial features of the problem. So ... when we look at the theory of condensed matter physics nowadays we inevitably talk about a 'model'. [And] a good model is like a good caricature: it should emphasize those features which are most important and should downplay the inessential details. (Fisher 1983: 47)

We are thus led to a conclusion parallel to Bas van Fraassen's rejection of the principle of inference to the best explanation and Nancy Cartwright's observation that 'The truth doesn't explain much': from the fact that a model M provides the best explanation of the behaviour of a system S it does not follow that M is a faithful representation of S.

Like many other forms of representation, a model is best understood as an epistemic resource. We tap this resource by seeing what we can demonstrate with the model. This may involve demonstration in the seventeenth-century sense of the word—that is, mathematical proof. Lars Onsager, for example, used strict methods of mathematical proof to show, for the two-dimensional Ising model in the absence of an external field, that the quantity C (denoting specific heat) diverges logarithmically as $|T - T_C|$ approaches zero; in contrast, Domb and others had perforce to use approximate methods, often supplemented by computer techniques, to show the corresponding result (that $\alpha = 0.12$) for the three-dimensional version. Or, it may involve demonstration as the term is currently used. Bruce and Wallace used computer simulations to demonstrate, through a series of pictures, the effects of coarse-graining. In both cases the internal dynamic of the model serves an epistemic function: it enables us to know more, and understand more, about the physical systems we are studying.

6.2.2. Computer Simulation

In section 6.1.8, I drew a distinction between the use of computer techniques to perform calculations, on the one hand, and computer simulation, on the other. The distinction may not always be clear-cut, but paradigm cases will help to confirm it.

A paradigm case of the former is the use of step-by-step methods to chart the evolution of a system whose dynamics are governed by non-integrable differential equation.[49] Notoriously, in Newtonian gravitational theory, for instance, there is no analytic solution to the three-body problem, even though the differential equations that govern the bodies' motions are precisely defined for all configurations of the system. But, from a given configuration of, say, the Sun, Jupiter, and Saturn at time t, we can use these equations to approximate their configuration at time $t + \Delta t$, and then at time $t + 2\Delta t$, and so on. Reiteration of this process yields their configuration at any subsequent time $t + n \, \Delta t$. (The specification of a configuration here includes not only the positions of the bodies, but also their momenta.) These are just the kinds of calculations that computers can perform more speedily than human beings. Interpolation techniques can then be used to ensure that these approximate solutions fall within acceptable limits. The simplest requires the computer to check that, if it performs two calculations for the successive time intervals from t to $t + \Delta t/2$, and from $t + \Delta t/2$ to $t + \Delta t$, the result is sufficiently close to that obtained for the time interval from t to $t + \Delta t$, and, if it is not, to decrease the value of Δt until an adequate approximation is reached.

A paradigm case of computer simulation is cited by Fritz Rohrlich. It is designed to investigate what takes place at the atomic level when a gold plate is touched by the tip of a nickel pin.[50] To quote Rohrlich:

[T]he simulation assumes 5000 atoms in dynamic interaction, 8 layers of 450 atoms constituting the Au metal, and 1400 atoms the Ni tip. The latter are arranged in 6 layers of 200 atoms followed by one layer of 128 and one of 72 atoms forming the

---

[49] For discussions of differential equations in physics that do not have analytic solutions, see Humphreys 1991: 498–500 and Rohrlich 1991: 509–10.
[50] The simulation was originally described in Landman et al. 1990.

tip. Additional layers of static atoms are assumed both below the Au and above the Ni; their separation provides a measure of the distance during lowering and raising of the Ni tip. The inter-atomic interactions are modeled quantitatively by means of previously established techniques and are appropriate to a temperature of 300K. To this end the *differential equations of motion* were integrated by numerical integration in time steps of 3.05×10^{-15} sec. The tip is lowered at a rate of $\frac{1}{4}$ A per 500 time steps.   (Rohrlich 1991: 511–12; emphasis in the original)

Notice the thoroughly realist mode of description that Rohrlich employs. Aside from his references to the ways inter-atomic interactions are modelled and numerical methods are used to integrate the equations of motion, the passage reads like a description of an actual physical process—albeit one which our technology would find it hard to replicate precisely. The impression of a realistic description is heightened by Rohrlich's inclusion of several computer-generated illustrations of the process. As he comments (1991: 511): 'There is clearly a tendency to forget that these figures are the results of a computer simulation, of a calculation; they are not photographs of a material physical model.'

The mimetic function of a simulation is emphasized in the characterization of a computer simulation given by Stephan Hartmann: '*A simulation imitates one process by another process*' (1996: 83). In this definition, the term 'process' refers solely to some object or system whose state changes in time. If the simulation is run on a computer, it is called a *computer simulation* (Hartmann 1996: 83; emphasis in the original). Hartmann contrasts this definition with the 'working definition' proposed by Paul Humphreys (1991: 501): '*Working definition*: A computer simulation is any computer-implemented method for exploring the properties of mathematical models where analytic methods are unavailable.'

Of these two proposals, Hartmann's can clearly be accommodated within the DDI account of representation. Recall, for example, that the account describes a theoretical model as possessing an internal dynamic. The dynamic has an epistemic function: it enables us to draw conclusions about the behaviour of the model, and hence about the behaviour of its subject. But, though the verification of these conclusions may lie in the future, these conclusions are not exclusively about future events, or even events in the future relative to the specification of the model we start from. They can also concern properties of the model coexisting with its specified properties, or general patterns of its behaviour. That is to say, the epistemic dynamic of the model may, but need not, coincide with a temporal dynamic.

As befits a general account of modelling, the DDI account includes as a special case the kind of 'dynamic model' which, according to Hartmann (1996: 83), is involved in simulation, and which 'is designed to imitate the time-evolution of a system'. Whether Hartmann is right to restrict himself to dynamic models of this kind is another question. After all, the prime purpose of the Ising model simulations performed by Pearson et al. (see section 6.1.8) was to investigate the variation of the model's properties with temperature, rather than with time. However, I will not press the issue here.

I have a more serious reservation about Humphreys' 'working definition', to wit, that it blurs the distinction I have just made between computer simulation and the use of the computer to solve intractable equations. This distinction appears very clearly within the DDI account of modelling and representation. As I emphasized, the dynamic of a model enables us to demonstrate conclusions. Very often this involves solving equations, and one way to do so is by the use of computer techniques, as in the case of the three-body problem. (Notice that it is not the only way. Newton was the first of a distinguished line of astronomers to apply perturbation theory to such problems.) Alternatively, however, we may choose to remodel the original representation using a computer—in other words, to nest one representation within another in the manner shown in Figure 5.4. The result is a hierarchy of models of the sort mentioned in section 6.2.1, save that in this case the whole of the second stage of the modelling process is performed on a computer. In this case, if the process involved in Demonstration 2 (see Figure 5.4) is a temporal process, then it conforms precisely to Hartmann's characterization of a computer simulation; it imitates the behaviour of the model.

All computer simulations of the Ising model are of this general kind. They also support one of Hartmann's criticisms (1996: 84) of Humphreys' 'working definition'. In denying Humphreys' suggestion that computer simulations are used only when analytic solutions are unavailable, Hartmann could well have pointed to the two-dimensional Ising model; computer simulations of its behaviour are used despite the fact that an exact solution already exists.

Two further points need to be made about the way computer simulation fits within the DDI account of representation. The first is that while the distinction between computer simulation and the use of computer-driven calculational techniques is conceptually clear-cut, individual cases may be hard to classify, as will appear in section 6.2.4. The second is that the two alternatives are not exhaustive. Common to my description of both of them was the assumption that the phenomena under study were represented by an abstract mathematical model (Model$_1$ in Fig. 5.4). What was at issue was simply the way the behaviour of the model was to be investigated: in one case, computer-aided computation was to be used; in the other, computer simulation. The assumption may not always hold. A third possibility is that the computer simulation can be a direct representation, so to speak, of a physical phenomenon. At a first level of analysis, for instance, the computer simulation discussed by Rohrlich, of a pin touching a plate, is a direct representation of just this kind. I use the phrase 'at a first level', because a more fine-grained analysis is possible. In the first place, a prepared description of the pin-plate system is used. The tip of the pin is described as a sequence of layers of Ni atoms, each containing a certain number of atoms. For this reason we may deny that the simulation directly represents the phenomenon. In the second place, even within the simulation itself the inter-atomic forces are modelled in a particular way, and a specific technique of integrating the relevant differential equations is used. Whereas the latter involves the use of a computer to perform a calculation, the former involves the nesting of one computer simulation within another.

A virtue of the DDI account of representation is that it allows us to articulate the complexity of computer simulation, a complexity over and above those encountered in the writing of the computer program itself. Of course, whether we need an analysis as fine-grained (not to say 'picky') as the one I have just offered will depend on the kinds of question we wish to answer.

6.2.3. Cellular Automata and the Ising Model

Since the seventeenth century, developments in physics have gone hand in hand with advances in mathematics. A theoretical model would be of no use to us if we lacked the requisite mathematical or computational techniques for investigating its behaviour. By the same token, it is hardly surprising that the arrival of the digital computer, a mathematical tool of enormous power and versatility, but at the same time a tool of a rather special kind, has stimulated the use of theoretical models that are particularly suited to computer simulation. One such model is the cellular automaton.[51]

---

[51] The study of cellular automata was initiated by John von Neumann in the late 1940s, with input from Stan Ulam. See Burke 1970. Von Neumann's original project known as 'Life' was to

A cellular automaton (CA) consists of a regular lattice of spatial cells, each of which is in one of a finite number of states. A specification of the state of each cell at time t gives the *configuration* of the CA at that time. The states of the cells evolve in discrete time steps according to uniform deterministic rules; the state of a given cell I at time $t + 1$ is a function of the states of the neighbouring cells at time t (and sometimes of the state of cell I at time t as well). In summary, space, time, and all dynamical variables are discrete, and the evolution of these variables is governed by local and deterministic rules. (As we shall see, the last stipulation is sometimes relaxed to allow probabilistic rules.) We may visualize a CA as an extended checkerboard, whose squares are of various different colours, and change colour together (some, perhaps, remaining unchanged at regular intervals).

The discreteness that characterizes cellular automata makes them ideal for computer simulation. They provide exactly computable models. Indeed, a realization of a CA is itself a computing device, albeit one with a rather limited repertoire (see Toffoli 1984: 120). The dynamic rules apply to all cells simultaneously, and so the CA enjoys the advantages of parallel processing. To put this another way, if we set out to design a computer that minimizes the energy expended in the wiring and the time that signals spend there, rather than in the active processing elements of the computer, we will be led towards the architecture of a cellular automaton.[52]

For the present, however, I will retain the distinction between a cellular automaton, regarded as a model of a physical system, and the computer simulation of its behaviour.[53] As models, cellular automata are radically different from those traditionally used in physics. They are exactly computable; no equations need to be solved and no numerical calculations need to be rounded off (Vichniac 1984: 118). Demonstration in these models does not proceed in the traditional way, by the use of derivation rules appropriate

---

model biological processes; the best-known cellular automaton was devised by John Conway in 1970, and described by Martin Gardner in a series of *Scientific American* articles in 1970 and 1971. See Gardner 1983. In 1984, *Physica*, 101, edited by Stephen Wolfram, brought together 'results, methods, and applications of cellular automata from mathematics, physics, chemistry, biology, and computer science' (viii). I draw on these essays in what follows.

[52] A nice example of a cellular automaton as a model, cited by Rohrlich (1991: 513–14), is a model on the evolution of a spiral galaxy, in which the galaxy is represented in a disc formed by concentric rings of cells. The hypothesis modelled is that if a star is born in a particular cell there is a probability p that another will be born in an adjacent cell within a time interval $\Delta(= 0.01$ billion years)!

[53] See Vichniac 1984: 96 and Toffoli 1984: 120, but bear in mind when these essays were written.

to the formulae of symbolic representations. Instead, only simple algebraic operations are required (Toffoli 1984: 118). Rohrlich (1991: 515) speaks of the 'new syntax' for physical theory that cellular automata provide, in place of the differential equations characteristic, not only of classical physics, but of quantum theory as well. He suggests that CAs will increasingly come to be used as phenomenological (as opposed to foundational) models of complex systems.

There are obvious structural similarities between cellular automata and the Ising model. And, given the theoretical significance of that model, the question naturally arises: Can the behaviour of the Ising model be simulated by a CA? As we saw in section 6.1.8, the task facing the automaton would be that of simulating the wandering of the Ising configuration through the canonical ensemble, so that the frequencies of occurrence of individual arrangements would accord with the probabilities given by Boltzmann's law (equation 6.3). A simple argument, due to Vichniac (1984: 105–6), shows that this cannot be done by a CA that uses fully parallel processing—that is, that applies its dynamic rules to all cells simultaneously.

The argument runs as follows. Consider a one-dimensional Ising model in a fully ordered spin-up arrangement, where, for every site I, $s_i(a) = +1$. Diagrammatically, we have:

$$\ldots + + + + + + + + + + + \ldots$$

Assume that the Ising model acts as a CA—that is, that each site I can be regarded as a cell, and that the value of $s_i(a)$ denotes its state. Let Ri denote the right-hand neighbour of site I, and Li its left-hand neighbour. In this arrangement, for each site I, $s_{Li}(a) = s_{Ri}(a) = +1$. It follows that, if the CA is fully deterministic, then either all sites will flip at every time step, or none will. This behaviour obtains in the Ising model only at absolute zero, when no transitions occur. Hence, at any other temperature T the CA must behave probabilistically. Since, however, we are still considering the wholly ordered arrangement, the value of the probability function involved will be the same at each site—that is, each site will be ascribed the same probability $p(T)$ of remaining $+1$, and $1 - p(T)$ of flipping to -1.

Assume, for the sake of argument, that the temperature of the lattice is such that $p(T) = 2/3$. Since N is very large, there is a non-negligible probability that a single transition will leave the lattice in an arrangement whereby, for two thirds of the sites, $s_i(a) = +1$, and for the remaining third,

$s_i(a) = -1$. Among the possible arrangements of this kind are these (in which the periodicity shown is assumed to be repeated throughout the lattice):

$$\ldots + + - + + - + + - + + - \ldots$$
$$\ldots + + + + - - + + + + - - \ldots$$
$$\ldots + + + + + + + + - - - - \ldots$$

Furthermore, each of these arrangements will be equiprobable.

They will not, however, all have the same total interaction energy, E_a, since each has twice as many domain walls (where neighbouring sites are of opposite sign as the one below). But, it is the value of E_a, rather than the proportion of sites where $s_i(a) = +1$, which, by Boltzmann's law, gives the probability of the arrangement a at a given temperature.

The argument, which is not restricted to the one-dimensional case, suggests that the reason why this simple CA simulation of the Ising model gives 'terrible results'[54] is that the rule governing the changes of state of cells of a CA are purely local; each cell's behaviour is governed by the states of its immediate neighbours. While rules could be formulated which reflected the nearest-neighbour interaction energies of the Ising model, no set of CA rules can be sensitive to the *total* interaction energy of the model of the system, as must be the case if the arrangements are to form a canonical ensemble.

We are led here to a deep question in thermodynamics: If interaction energies are due to local interactions, as in the Ising model, whence comes the dependence of the probability of a particular arrangement on a global property, the total interaction energy? I will not try to answer it. Note, however, that an adequate simulation of the Ising model's behaviour can be achieved by abandoning full parallelism, and using either (1) the site-by-site Monte Carlo method described in section 6.1.8, or (2) a CA, modified so that its rules are applied alternately to every second cell (Vichniac 1984: 104). This suggests that the non-divisibility of time in the simple CA simulation may bear part of the blame for its failure to imitate the Ising model.[55]

This thought leads in turn to a general comment about the use of cellular automata as models of physical systems. As we have seen, they have been

---

[54] Vichniac reports that 'a spurious checkerboard pattern starts to grow, leading eventually to a state of *maximum* energy' (1984: 105; emphasis in the original).

[55] For further discussion, see Vichniac 1984: sect. 4.5.

hailed as providing 'an alternative to differential equations' (Toffoli 1984), and 'a new syntax' for physical theory (Rohrlich 1991). Vichniac declares (1984: 97):

They provide a third alternative to the classical dichotomy between models that are solvable exactly (by analytic means) but are very stylized, and models that are more realistic but can be solved approximately (by numerical means) only. In fact cellular automata have enough expressive power to represent phenomena of arbitrary complexity, and at the same time they can be simulated exactly by concrete computational means. ... In other words, we have here a third class of *exactly computable* models.

Amidst all this enthusiasm, far be it for me to play the role of the designated mourner. But, as the example of the Ising model shows, a word of caution is called for.[56] The reason is straightforward. The constraints that prepared descriptions of physical systems must conform to before they become eligible for modelling by cellular automata are severe. That is the price paid for the exactness of the computations that these devices provide. I have no doubt that physical systems exist that decline to be squeezed into this straitjacket; systems, for example, whose dynamics depend crucially on the continuity, or at least the density, of the time series. For such a system, a cellular automaton would not offer a reliable representation.

6.2.4. Computer Experiments

In an experiment, as that term is generally used, an experimenter interacts with some part of world in the expectation of learning something about it.
A computer is 'a device or machine capable of accepting information, applying prescribed processes to the information, and supplying the results of those processes'. (I quote Van Nostrand's *Scientific Encyclopedia*, 5th edn (1968).)

So, when physicists talk of 'running experiments on the computer', what do they mean? Not, presumably, that these experiments are to be performed in order to learn something about computers. But, if not, then what else? I will approach the question in four steps, which the reader may well regard as a slippery slope. Each step will involve an example.

Observe, first of all, that not all of the models used in physics are abstract mathematical models. Material models, like the scale models used

---

[56] It appears at Longair 1989: 185.

in wind tunnels, are also used. Joseph Smeaton's work on waterwheels in the eighteenth century provides a historical example. In the 1750s, Smeaton compared the advantages of overshot and undershot wheels by experimenting with a model waterwheel, in which the relevant parameters could be readily adjusted. As Davis Baird remarks:

Smeaton's waterwheel is not, however, a scaled down version of some particular waterwheel he hoped to build. His model serves the more abstract purpose of allowing him to better understand how waterwheels in general extract power from water in motion. ... In quite specific ways it functioned just as theory functions. (Baird 1995: 446)

The dynamic of the model, in both the mechanical sense and the epistemic, was supplied by a physical process, the same process, in fact, that is at work in full-sized waterwheels. By using this dynamic, Smeaton drew conclusions about a waterwheel's behaviour. More precisely, he demonstrated conclusions for the model wheel, which he assumed would apply to a full-sized version. In this enterprise, Smeaton was clearly conducting experiments, albeit experiments on a *model* waterwheel.

The second example comes from one of Richard Feynman's *Lectures on Physics* (Feynman, Leighton, and Sands 1963–5: i. 25.8), in which these two equations are displayed:

$$m(d^2x/dt^2) + \gamma m(dx/dt) + kx = F$$
$$L(d^2q/dt^2) + R(dq/dt) + q/C = V$$

Both equations describe damped forced oscillations. In the upper equation, where x and t denote distance and time respectively, the oscillations are those of a mechanical system like a pendulum when an external force is applied; m is the mass of the system, and F the external force, which is assumed to vary sinusoidally with time. The oscillations are damped by a frictional force that varies directly with the speed of the mass: $F_f = \gamma m(dx/dt)$. The factor k determines the natural frequency n of oscillation of the system according to the formula $n = \sqrt{k}/2\pi$. The lower equation also describes oscillations, but they are the electrical oscillations of a circuit under a sinusoidally varying voltage V. Here L is the self-inductance of the circuit, R its electrical resistance, and C its capacitance. Since q denotes electric charge, dq/dt denotes electric current, and d^2q/dt^2 the rate of change of current.

The two equations are in exact correspondence with each other (but notice that R corresponds to γm, and $1/C$ to k). It follows that we can investigate the behaviour of any mechanical system subject to damped forced oscillations by looking at the corresponding electrical circuit, not by solving its equations, which will be the same for both systems, but by building the electrical circuit and establishing its behaviour by experiment. Feynman adds:

This [the electrical circuit] is called an *analog computer*. It is a device which imitates the problem that we want to solve by making another problem, which has the same equation, but in another circumstance of nature, and which is easier to build, to measure, to adjust, and to destroy! (ibid.)

As in the first example, this procedure clearly involves experimentation— experimentation on a computer, to boot. The added factor in this example is that the results of these experiments are interpreted in terms of a totally different type of system. The physical processes at work in the electrical circuit, and which supply the epistemic dynamic of the representation, are not the same as those at work in the mechanical system.

The third example involves the cellular automaton. As I pointed out, a realization of a cellular automaton (a *cellular automaton machine* or *CA machine*) is itself a computer, and can be used to provide simulations of a physical system's behaviour. But, unlike an ordinary computer, a cellular automaton is tailored towards a specific system; the structure of the automaton is designed to match crucial structural properties of the system being modelled, like its topology and its symmetries. Since a CA machine is a physical system, its behaviour can be the subject of experiments. Some of these experiments—those that explore the properties that the CA system has in common with the modelled system—will yield conclusions that can be interpreted in terms of that system.

In the first two examples, the idea of performing experiments on one system to obtain information about another is unexceptionable, and in the third it does not seem far-fetched. In the fourth and final example, however, it may seem problematic. For this example, I return to the computer simulation of atomic interactions between a pin and a plate (section 6.2.2). Would we say that Landman and his collaborators were *conducting experiments*? If not, I suggest, the reason is this. As in the three earlier examples, their simulation made use of a material object, in this instance a highly powered computer. And in order to make the tip of the image of the pin descend towards the image of the plate at a certain speed, various processes were set in train within the computer. But, whereas in the cases of the model waterwheel, the electrical circuit, and the cellular automaton there was an obvious correspondence between the processes at work in the simulating device and those at work in the system being modelled, in the present case no obvious correspondence exists. Indeed, that is the reason why the all-purpose, high-speed computer is so versatile; its design enables it to execute a great variety of abstract programs. The internal processes within such a computer are governed by physical laws, but different computers will apply different laws or the same laws in different ways, to perform the same tasks. To use a cliché, the same software can be implemented in many different kinds of hardware. Thus, Landman's methodology differed from Smeaton's in this respect, that Smeaton investigated the properties of a full-sized waterwheel by examining the properties of a model waterwheel, but Landman did not investigate the dynamics of pin-and-plate interactions by examining the dynamics of a computer.

None the less, like the systems examined in the three earlier examples, the computer is a material system, and in using it to provide answers to our questions we are relying on the physical processes at work within it. If we ask what gives rise to, or what licenses expressions like 'running experiments on the computer', the analogies between the four examples are part of the answer. Another is the degree to which, in the late twentieth century, the lived world of physicists who perform orthodox experiments conforms to the world of those who do their experiments on a computer. The orthodox experimentalist works surrounded by electronic devices for enhancing signal at the expense of noise, for sharpening images or smoothing out irregularities, for performing statistical analyses of data collected, etc., etc.

Consider, as an example, the displaying of results. On the desk in front of me as I write (literally) is a book, open to reveal a striking but enigmatic picture. Its central panel could be a reproduction of an abstract painting. It is a vividly coloured rectangle, five inches wide and half as high. No fewer than sixteen different colours are present in it, ranging from black and dark blue at the lower left-hand side to bright red and white near its upper right-hand corner. Separating these areas, a broad river of intermediate colours, blues and greens below, yellows and oranges above, meanders across the page, as though tubes of non-miscible ink had been squeezed out to produce it. The colours nowhere blend into one another, and the sharpness with which they are defined suggests that the picture has been prepared on a computer. That impression is confirmed by the shape of the broad black border that surrounds the panel, and gives the complete picture the proportions of a television screen.

A heading 'SPACE RESEARCH INSTITUTE, USSR' appears in the upper section of the border, and in the lower section there is a horizontal strip in which the spectrum of colours from the panel are displayed, with the black end labelled '−3.5', a light-green segment in the centre, '.5', and the white end, '4.5 MK'. From these clues we can infer that the panel presents the results of experiment or observation. But, if the former, are they the computerized results of an experiment, or the results of an experiment on a computer? Even the legend 'RADIO MAP' below the panel and the coordinates running along its bottom edge and up one side are not conclusive. There are well-known maps of Treasure Island and Middle Earth.

In fact, the picture presents observations made by the Russian spacecraft RELIKT 1. The coloured panel is a projection onto the plane of the Galactic Sphere, a celestial sphere in which coordinates are measured from the galactic plane rather than the plane of the Earth's equator. Recorded on it are the intensities of the Microwave Background Radiation (the echo of the Big Bang) from different regions of the sky at a wavelength of 8 mm; hence the phrase, 'Radio Map'. But, what we see on the page are not the 'raw data' of these observations; they have been carefully processed for our consumption:

The image is a 'differential map' in the sense that all the intensities are measured relative to the intensity of the Microwave Background Radiation at a fixed point in the sky which has brightness temperature 2.75K. The colour coding of the intensity corresponds to a temperature difference between the maximum and minimum of +8mK and −4.5mK respectively.   (Longair 1989: 96)

The image has also been 'smoothed to a beam width of 20°' (ibid. 185).

In section 6.2.2, I commented on the sense of realism that visual presentations impart to computer simulations. Ironically, this sense of realism is reinforced, if not induced, by the manifest artificiality with which the results of orthodox experiments are presented. Both confirm us in our use of the expression, 'experiments on a computer'. None the less, however selective the sensors on the Russian spacecraft were, however vigorously the signals from them were massaged, and however arbitrary was the colour-coding of the final picture, the distribution of colours on the page before me is due in large part to the interactions between the sensors and the external world. Indeed, the function of the commentary that accompanies the picture, and tells us what procedures were used in its production, is precisely to help us understand what kind of information it provides.

This prompts the question: What kind of information can be obtained from computer experiments, investigations in which the only interaction is between the experimentalist and the computer? Well, it can be information about the actual world, about possible worlds, or about impossible worlds. For the first, paradoxical, case, let us return, for the final time, to the two-dimensional Ising model. Amongst the computer experiments performed by Bruce and Wallace (see section 6.1.7), the behaviour at the critical point of two versions of the model was compared. In the first version the site variable took one of two values (-1 or $+1$); in the second it took one of three values (-1, 0, or $+1$). That experiment showed that, whereas the local configurations for the two versions were quite different, their configuration spectra both approached the same limit under coarse-graining. This result neatly illustrates the fact that critical behaviour at large-length scales may be independent of seemingly important features of the atomic processes that give rise to it. It acquires significance through the importance of the Ising model in our understanding of critical phenomena. Because the Ising model acts as a representative, as well as a representation, of a whole universality class of systems, the experiment tells us something about the actual world.

Examples of computer experiments that provide information about a possible world are not far to seek. One such would be a simulation to determine the effects on the Earth's orbit, were a heavy asteroid to come close to it. Conceptually more interesting, however, are experiments to investigate the behaviour of physical entities that may or may not exist. It is, for example, possible that cosmic 'defects'—persistent inhomogeneities of the Higgs field, analogous to magnetic domain walls—existed in the early universe. If so, they may give rise to fluctuations in the Microwave Background Radiation that is our legacy from that epoch. Alternatively, such fluctuations may be due to an early period of cosmic inflation. The effects of defects may be represented by adding additional terms to the equations that, according to our best theory, governed pressure waves in the early universe, and computer experimentation makes it possible to show that the differences between the predictions of the two models are large enough to be registered by the new generation of satellite-based instruments. They 'translate into observable differences in the micro-wave sky' (Albrecht et al. 1996: 1413). Thus, these computer experiments tell us about two possible worlds, one of which may happen to be our own.

A computer experiment that gives us information about impossible worlds is one that investigate the effects of changes in the fundamental constants of physics or of its fundamental laws, since those constants and those laws define the notion of physical possibility. Such experiments have been performed by cosmologists, and seized on by proponents of the anthropic principle, for whom a draught of Hume 1776, twice daily, is recommended.

The fact that computer experimentation can give us information about many kinds of worlds, however, tells us something that should have been obvious all along: that, lacking other data, we can never evaluate the information that these experiments provide. We know that the computer simulation of the Ising model gives us information about the actual world, because we have independent evidence of the model's significance; we will know whether or not the theory of cosmic defects is adequate, not via computer experiments, but through the use of satellite-based instruments. In other words, though the use of computer experimentation is in a crucial respect on a par with all other kinds of theoretical speculation, the difference between them lies in the richness and variety of the information that the computer provides about the theoretical consequences of those speculations. Of course, doing 'imaginary physics', charting the physics of hypothetical worlds, may bring its own rewards. As Giuseppe Verdi said, it is a fine thing to copy reality, but a better thing to invent it.

7

Theoretical Explanation

> Explaining metaphysics to the nation—
> I wish he would explain his explanation.
>
> Lord Byron

DESIDERATA

By a *theoretical explanation*, I mean an explanation of a regularity in the world, or in our account of the world, in terms of a scientific theory. An account of theoretical explanation should satisfy various desiderata. For example:

- The account should be recognizable as a part of a broader picture of explanation.
- It is should also be linked to a general account of the structure of scientific theory.
- It should be descriptively accurate; it should accommodate anything we would normally consider a theoretical explanation.
- It should allow for the evaluation of explanations; it should show where, and on what grounds, such evaluations are made.
- It should be sensitive to the achievements of earlier accounts of scientific explanation, to show how they succeeded, and where they fell short.

These desiderata have guided the account of theoretical explanation I offer here. In capsule form, it runs as follows. We explain some feature X of the world by displaying a theoretical model M of part of the world, and demonstrating that there is a feature Y of the model that corresponds to X, and is not explicit in the definition of M.[1]

---

[1] For a historical account of work on the topic of scientific explanation, see Salmon 1989.

7.1. EXPLANATIONS

All manner of things require explanation. We can ask someone to explain to us what made the Beatles pre-eminent, what distinguishes analytic cubism from synthetic, how to make an omelette, how the West was won, why giraffes have long necks, why the seven stars are no more that seven, and so on and so forth.

We make such requests in the hope of increasing our understanding. The explanation may also help us to acquire a skill, like the ability to make an omelette, but, in asking for an explanation rather than a demonstration, we are seeking an increased cognitive awareness of the process involved. Acquiring the skill may involve breaking a few eggs.

To provide an explanation is often to answer a question, as my examples show. Contrary to standard views, however, this is not always the case. Non-native speakers of soccer sometimes ask me to explain the offside rule to them. I may be self-indulgent on these occasions, but there is no individual question to which my response is the answer. Likewise, if I were asked to explain the periodic table of the elements or the special theory of relativity, either request would seem perfectly legitimate, but the result would not be reducible to a cluster of answers to bare what-, how-, or why-questions.

Of course, some explanations of general topics provide answers to questions. Newton's explanation of the rainbow answers the question, 'How is the rainbow formed?' But Newton himself describes his project very generally as, 'By the discovered Properties of Light to explain the Colours of the Rainbow' ([1730] 1952: 168). His explanation answers, not just a single question, but a large and not clearly specifiable family of questions. Consider Cotton Mather's encomium on Newton's achievement, written in 1712:

This rare Person, in his incomparable Opticks, has yet further explained the Phenomena of the Rainbow; and has not only shown how the Bow is made, but how the Colours (whereof Antiquity made but three) are formed; how the Rays do strike our Sense with the Colours in the Order which is required by their Degree of Refrangibility, in their Progress from the inside of the Bow to the Outside: the Violet, the Indigo, the Blue, the Green, the Yellow, the Orange, the Red.

Mather distinguishes the question, 'how the Bow is made', from various others to which Newton provided the answers; he could have cited still others, all embraced by Newton's explanation of the rainbow. My own use of the phrase 'the explanation of the rainbow' is not idiosyncratic. Carl Boyer, in his book on the rainbow from which the quotation is taken, talks of Kepler 'believing that the explanation of the rainbow would be expedited ...' ([1959] 1987: 183), of Newton 'beginning his explanation of the rainbow' (ibid. 252), and of Airy, Mills, and Potter producing 'the definitive explanation of the rainbow' (ibid. 294). Nor is this list exhaustive.

I return to the rainbow, if not the offside rule, in section 7.3. Setting aside for now explanations of general topics like these, we may agree that many, perhaps most, explanations provide answers to questions. These questions may be of various kinds; the examples listed earlier all began with the interrogatives 'what', 'how', or 'why', but other openings are possible.[2] Among philosophers of science, however, there has for half a century been a consensus verging on unanimity, not only that all scientific explanations provide answers to questions, but that the only questions they answer are why-questions.[3]

Why should we believe this? Look again at Cotton Mather's tribute to Newton. In the list of questions to which Newton provided answers, there is not a why-question to be found. All are how-questions. Were Newton's explanations of these phenomena not properly scientific?

One might argue that to every how-question that is a request for a scientific explanation there corresponds a synonymous why-question. This would be a bad response, for two reasons. The first is that it is not obviously true. Maybe the question, 'Why do rainbows occur?' is the same question as Mather's, 'How is the Bow made?' but there is no why-question that naturally corresponds to 'How do the kidneys clean the blood?' Certainly it is not 'Why do the kidneys clean the blood?' The second reason is that, even if we find a means of translating how-questions into why-questions, the result may mislead us, since the logic of why-questions is not the logic of how-questions. If we consider why-questions on their own, then it may seem obvious that every why-question presupposes a contrast class, and that implicit in 'Why this?' is the phrase, 'rather than that'. We may go on to build this fact into our theory of explanation.[4] But 'How this, rather than that?' is a very peculiar construction indeed; there is no plausible contrast class that we can associate with, 'How do the kidneys clean the blood?' The price paid for

---

[2] Sylvain Bromberger (1965: 74) suggests 'whence' and 'whither'.
[3] For instance, van Fraassen 1980: 134: 'Any explanation is an answer to a why-question.' Salmon 1984: 'A request for a scientific explanation ... can always be reasonably posed by means of a why-question.'   [4] As does van Fraassen 1980: 141–2.

apparently minor artificialities of translation is a skewed account of scientific explanation.[5]

As my title suggests, this essay deals only with a limited class of explanations. This class, however, contains answers to how-questions as well as why-questions.[6] Even so, there are many scientific explanations that fall outside it. In the first place, I confine myself to what it is for a *theory* to explain some feature X of the world; secondly, I consider only cases where X is a *general regularity*, rather than the occurrence of a specific event. Each of these restrictions requires some comment, starting with the restriction to theoretical explanations.

Even with those restrictions, the subject of the verb 'explain' can be a state of affairs, a theory, or a person:

- The difference between the moon's gravitation tug on the oceans on each side of the Earth, together with the diurnal rotation of the Earth, explains the tides.
- Newton's theory of gravitation explains the tides.
- Newton explains the tides.

An analysis by Peter Achinstein (1983) favours the third option. For him, *explaining* is a speech act; more precisely, it is an *illocutionary* act. Achinstein proposes (C), below, as the fundamental condition necessary for a speaker S to have performed the act of explaining q. Perhaps surprisingly, the condition makes no reference to the intended audience.

(C) S utters u with the intention that his utterance of u render q understandable.
(ibid. 16)

On Achinstein's account, the act of explaining is logically prior to the concept of an explanation: a given proposition is only an explanation if it is expressed in an illocutionary act of explaining.[7] He calls this the 'ordered pair' view

---

[5] On this issue, I was awakened from my dogmatic slumbers by a paper by Maria Trumpler 1989; her examples came from recent work in physiology on how the nerves transmit information.

[6] Thus, my project is very different from Wesley Salmon's *Scientific Explanation and the Causal Structure of the World* (1984).

[7] Achinstein uses the taxonomy of speech acts laid out by J. L. Austin in *How to Do Things with Words* (1965), where the latter shows how the utterance 'Shoot her' can be described as three different kinds of speech acts by the person to whom the utterance is directed (ibid. 101–2):

Act A or *Locution*
He said to me 'Shoot her!' meaning by 'shoot' *shoot*, and by referring by 'her' to *her*.
Act B or Illocution

(ibid. 83–8): an explanation is an ordered pair of the form {proposition, act of explaining}. His account emphasizes the pragmatic nature of explaining and explanation,[8] and with some important reservations, I am sympathetic to the view. It does, however, raise an immediate difficulty for my project, in that I talk of a theory explaining X, but would never claim that a theory performed a speech act.

This particular difficulty is more apparent than real. Even if we adopt Achinstein's approach, we may still enquire what kinds of utterances can be the first member of an explanation. Thus, we can regard 'Theory T explains X' as an ellipsis for, 'A speaker could use the resources of T to explain X'. My project is then to unpack the phrase 'using the resources of T'. For reasons that will appear in section 7.2, I prefer that mode of expression to one that speaks in terms of the *propositional content* of T.

A problem for any account of explanation is the *problem of audience preparedness*: given the differences in acumen and background knowledge from one listener to another, can one say anything useful about making a putative explanation understandable? For example, we may wonder what proportion of the audience understood Einstein when, on a autumn day in 1915, he presented his paper, 'Explanation (*Erklärung*) of the Perihelion Motion of Mercury by the General Theory of Relativity', to the Royal Prussian Academy of the Sciences. Hempel sidestepped the diversity problem by denying that there was a place in the logic of explanations for pragmatic notions like comprehensibility, which varied from one individual to the next.[9] In a sense, Achinstein has a different answer to the problem. For him, explaining involves only the intention to make something understandable; as I have noted, his fundamental condition makes no reference to the audience. But, should we accept the claim that explaining is unequivocally an illocutionary, rather than a perlocutionary act? The mark of a perlocutionary act, like persuading or convincing, is that it brings about an

> He urged or advised, ordered, &c. me to shoot her.
> Act C*a* or Perlocution
> He persuaded me to shoot her.
> Act C*b*
> He got me to or made me, &c. shoot her.

These four examples are followed by a matching quartet of *Locution, Illocution,* and two *Perlocutions,* centred on the utterance, 'You can't do that'.

[8] In contrast, in the Hempel–Oppenheim view, an explanation was either a syntactic or a syntactic-*cum*-semantic affair, depending on one's view of entailment.

[9] See Hempel 1965: 413. For another response to this problem, see Friedman 1988: 5–6.

effect on the listener (see n. 7). Just as we make unsuccessful attempts to persuade, we may make unsuccessful attempts to explain, and our failure may be due, not to ineptitude on our part, but to the obduracy of the listener in the first instance, and his obtuseness in the second. Or so we may think. Note that even on Achinstein's account the audience cannot be utterly ignored, if only because we cannot intend what we know to be impossible.

In practice, the problem of audience preparedness can be meliorated, at least in part, by two considerations. Consider first two explanations of, say, the Aharanov–Bohm effect, one appearing in *Scientific American* and the other appearing in J. J. Sakurai's *Advanced Quantum Mechanics*; the explanations will differ simply because the two publications have different readerships, each with is own commonality of interests and shared background knowledge, as the authors (and editors) of the two explanations know. The second is that, although within a given readership the preparedness of individuals may differ, a general account of explanation can accommodate those differences by noting that 'understandable' can be adverbially modified. The explanation appearing in *Scientific American*, for instance, will make the A–B effect completely understandable to some readers, barely understandable to others, and something in between to the majority.

I turn now briefly to the second restriction on my account. My reasons for examining explanations of general regularities rather than particular events are the same as Michael Friedman's. In his essay, 'Explanation and Scientific Understanding' (1988: 5–19), he points out that, while in scientific literature explanations of general regularities are much more frequent than explanations of particular events, in philosophical literature the situation is the reverse.

7.2. REPRESENTATION

A regularity in the physical world implies the existence of a physical system, perhaps a two-body system like Mercury and the Sun, or a naturally occurring conjunction like that of the Sun and rain, or a plasma in a metal in the solid state, or a hydrogen atom, or a nucleus. The task of the theoretician is to provide a representation of the behaviour of the system under scrutiny and in so doing to explain it. The representation may take various forms. In 1704, Newton used the laws of optics, specifically those of reflection and refraction, to explain how rainbows are brought about by the interaction between sunlight and raindrops. In 1913, Niels Bohr put forward a model for the hydrogen atom to account for the Balmer series in the spectrum of hydrogen. In 1915, Einstein applied his general theory of relativity to resolve a puzzle involving Mercury's orbit around the Sun; his representation of the Sun–Mercury system would now be called a *model*, as I pointed out in Essay 1. In the 1940s and 50s, two sub-disciplines of physics emerged, solid-state physics and nuclear physics, together with the models they employed. Many of those have figured in previous Essays.

The importance of models for solid-state physics was brought out by a paragraph, quoted earlier in Essay 2, in which the physicist David Pines describes a plasma within a metal as 'a gas containing a very high density of electrons and ions'. He wrote, in retrospect:[10]

In any approach to understanding the behaviour of complex systems, the theorist must begin by choosing a simple, yet realistic model for the behaviour of the system in which he is interested. Two models are commonly taken to represent the behaviour of plasmas. In the first, the plasma is assumed to be fully ionized gas; in other words, as being made up of electrons and positive ions of a single atomic species. The model is realistic for experimental situations in which the neutral atoms and impurity ions, present in all laboratory plasmas, play a negligible role. The second model is still simpler; in it the discrete nature of the positive ions is neglected altogether. The plasma is thus regarded as a collection of electrons moving in a background of uniform positive charge. Such a model can obviously only teach us about electronic behaviour in plasmas. It may be expected to account for experiments conducted under circumstances such that the electrons do not distinguish between the model, in which they interact with the uniform charge, and the actual plasma, in which they interact with positive ions. We adopt in what follows as a model for the electronic behaviour of both classical plasma, and the quantum plasma formed by electrons in models.   (Pines 1987: 65)

Although Pines employs neither 'explain' nor 'explanation', this paragraph is suffused with the concept of explanation:

In any approach to *understand* the behaviour of complex systems ... Such a model can obviously only *teach us* about electronic behaviour in plasmas. It may be expected to *account for* experiments.   (ibid.; emphasis added)

---

[10] 'In retrospect' because, in the text of the four papers which comprise the Bohm–Pines quartet, the term 'model' appears only in the fourth paper, written by Pines. He refers several times to 'an independent electron model' in the first two sections and nowhere else.

I call the models described by Pines *constitutive models*. Plasmas can be represented in two ways: first, as 'being made up of electrons and positive ions of a single atomic species'; secondly, as 'a collection of electrons moving in a background of uniform positive charge'.[11] In the last sentence of the paragraph he announces: 'We adopt [the second model] in what follows as a model for the electronic behaviour of both classical plasmas and the quantum plasma formed by electrons in solids.' In other words, he introduces another dimension to the model: more than foundational theories that can propel the deductions that drive the Bohm–Pines Quartet.

Around the time that Bohm and Pines pursued their investigations, a plethora of simple models of the atomic nucleus was emerging.[12] They included the liquid-drop model, the shell model, and the optical model.[13] Each of them was an *analogue model*, and each represented an aspect of the nucleus' behaviour. The liquid drop represented the system as a drop of incompressible 'nuclear liquid', and offered a theoretical account of nuclear fission; the shell model used a version of the Bohr atom, suitably scaled, to show why some nuclei are markedly less prone to decay than others; the optical model represented the nucleus as a homogeneous medium with a 'complex refractive index'; it provided an account of what happens when a high-energy neutron enters the nucleus.

Along with constitutive models and analogue models are models that mediate between foundational theories and the physical systems they represent, the 'centripetal forces' defined in the opening pages of Newton's *Principia*, for example, or the idealized and simplified model that represented the Sun-Mercury system in Einstein's 1915 paper.[14] These I call 'foundational models'.

7.3. THE RAINBOW

As a case study in explanation, I will use Newton's explanation of the rainbow. I choose this example because it is simple enough to be presented in full, yet

[11] At different times, Pines used each of these models, but in reverse chronological order. In the model in the Bohm–Pines quartet (1951–3), a collection of electrons was depicted as moving against a background of positive charge; in the model used in a subsequent paper (Pines 1955) the electrons moved within a regular lattice of positive ions.

[12] A footnote to the second paper of the Bohm–Pines quartet runs BP II: 339b: 'One of us D. B. and M. I. Ferenc are currently investigating a nuclear model based on a collective description of nucleon interactions.' [13] See Essay 4, section 4.1.4.

[14] The 'centripetal forces' are described in the model of the Sun-Mercury in Einstein's GTR (see Essay 1).

rich enough to illustrate a large number of issues. Though formulated in terms of an outdated theory of light, it is not just of historical interest. Versions of it still appear in high school textbooks, and I will compare Newton's presentation with the one offered by G. R. Noakes in a textbook 'written for the use of Sixth Form students [*British students in their last two years of high school*] who are specializing in science' (Noakes 1957: v), and with another appearing in *Physics*, a text devised in the 1960s by the Physical Sciences Study Committee (PSSC) on behalf of American students of the same age but with less exposure to mathematics than their European contemporaries.[15] A composite version runs as follows. I apologize if it reminds the reader of days best forgotten.

(*A*) Rainbows are formed when the Sun shines on falling rain. To observe them, the observer must face the rain with his back to the Sun. They appear because some of the Sun's rays enter the raindrops, are reflected within them, and then emerge towards the observer.

(*B*) Consider a ray from the Sun of a particular colour which strikes a spherical drop at the point B with angle of incidence i, and continues inside the drop with angle of refraction r (see Fig. 7.1). From Snell's law, $\sin i = \mu \sin r$, where μ is the refractive index (for Newton, the 'degree of refrangibility') from air to water for the light. The ray is internally reflected at D. The law of reflection, and the symmetrical geometry of the drop, together guarantee that, when the ray leaves the drop at E, the angle of incidence in the water–air surface is r. Since the trigonometric relationship expressed by Snell's law is independent of the direction of travel of the light, the angle of emergence is i. Thus, the ray suffers three deviations; it is refracted on entering the drop, reflected at the rear surface of the drop, and refracted again on leaving it. By simple geometry, the total deviation d is equal to $180° + 2i - 4r$.

(*C*) The incident ray depicted in the diagram is a representative instance of a group of rays, each ray parallel to the others. Thus, the angle of incidence is a variable; $i = 0°$ when the incident ray coincides with a diameter of the sphere, and $i = 90°$ when the incident ray is tangential to the sphere.

(*D*) Since r is obtainable from i by Snell's law, it follows that d is a function of i. It turns out that, as i moves from $0°$ to $90°$, the angle d decreases

---

[15] See, respectively, Newton [1730] 1952: 168–85; Noakes 1957: 116–20; Physical Sciences Study Committee 1960: 222–3. To make notation uniform, I quote these authors with minor changes in, e.g. the labelling of points in the diagram.

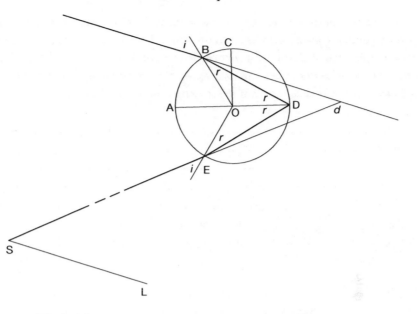

Fig. 7.1. The Rainbow.

from $180°$ to a minimum value of about $140°$, and then increases again to about $165°$. Elementary calculus shows that, at minimum deviation, cos. $i = \sqrt{\{(\mu^2 - 1)/3\}}$. For a wide range of angles of incidence, the deviation will be at or near this minimum value. Hence, we would expect the drop to reflect a lot of light in this particular direction. Conversely, we would expect a spectator looking at a rain shower to receive most light from those drops situated at places from which light reflected with minimum deviation would reach him.

We are now in a position to answer individual why-questions: Why do these drops form a bow? And why does the light appear coloured?

(*E*) To answer the first question, imagine a spectator receiving a ray of light reflected at minimum deviation from a drop. We may represent the spectator's eye by a point S on the emergent ray in Figure 7.1. The radius of the drop is very small compared with its distance from the spectator, so that SO is effectively collinear with the emergent ray. If we now draw a line SL through S parallel to the incident ray—that is, parallel with the Sun's rays, and then rotate the plane of the paper about the axis SL, then (a) the position of S will remain unchanged; (b) the line representing an incident ray will remain parallel to SL, and so can continue to represent a ray of light from the

Sun; and (c) the points O, O′, etc. will move in circles about the axis SL. Just those drops whose centres are represented by points on these circles will reflect light with minimum deviation towards the spectator. Those points (and there can be many of them) will lie on the surface of a cone with apex S; it follows that the spectator will see the corresponding drops as a bright bow in the sky.

(*F*) The colours of the bow are now easily explained. The refractive index μ for light going from air to water varies with the colour of the light. For red it is about 1.333, for violet light about 1.346. For this reason, the value of *d* at minimum deviation suffered by red light is about $2°$ more than that suffered by violet. Hence, 'the position of the drop determines not only whether it will contribute light to the rainbow but also the colour of its contribution' (Physical Sciences Study Committee 1960: 223). The result is a bow whose inner edge is violet, and whose outer edge is red.

For simplicity I have omitted from this account any discussion of the secondary and higher bows. These bows are produced by rays that have been reflected twice or more within the drop. The minimum deviation argument can be extended without difficulty to cover these bows, and, in fact, Newton and Noakes both do so.

The three authors differ, however, in the way they present the minimum deviation explanation. The most obvious differences appear in their presentations of the material of paragraph (*D*). Noakes, for instance, uses the calculus to prove that, at minimum deviation, *i* is given by $\cos i = \sqrt{\{(\mu^2 - 1)/3\}}$, and then adds, rather laconically (1953: 118): 'The value dd/di will be small in the neighbourhood of the minimum value of *d*, and thus for each colour the direction of minimum deviation will be that of greatest intensity.'

In contrast, Newton ([1730] 1953: 171) simply quotes the expression for *i* at minimum deviation, and assures his readers that 'The truth of all this Mathematicians will easily examine'. His explanation of why minimum deviation is associated with greatest intensity is, however, considerably more ample than Noakes'. He writes:

Now it is to be observed, that as when the Sun comes to his Tropicks [i.e. at the solstices], Days increase and decrease but a very little for a great while together; so when by increasing [the angle *i*] the Angles [of deviation] come to their Limits, they vary their quantity but very little for some time together and therefore a far greater number of the Rays which fall upon all the points in Quadrant AC, shall emerge in the Limits of their Angles than in any other Inclination.   (ibid.)

The PSSC makes no gesture towards a mathematical derivation of the minimum deviation result. Instead, they include a photograph of an experiment to show that the result is to be expected. A pencil of rays is passed through a hole in a screen, and strikes a small refracting sphere. The light reflected back to the screen illuminates a circular area, with most of the light concentrated at the outer edges of the circle. One may ask to what extent this demonstration is explanatory—especially when compared with the explanations offered by Noakes and Newton—and I shall return to this question in the next section.

Before discussing the further differences between the versions of this explanation, I will point out their common features. All three versions share the geometrical argument. A circle represents a raindrop, a straight line represents a light ray that travels towards a point B on the circle, and a geometrical construction by which two equal chords BD and DE represent light rays within the raindrop, and another straight line ES a light ray that travels from the raindrop to the spectator. Changes of direction of the ray at B, D, and E are governed at B and E by the laws of refraction, and at D by the laws of reflection. Absent in the diagram are a reflected ray at B, a refracted ray at D, and a reflected ray at E. The criterion for inclusion or exclusion is not the comparative intensities of the light rays, but whether the ray is germane to the argument or not. Indeed, the intensity of the refracted ray that exits from D may well exceed that of the reflected ray there. The PSSC version (1960: 223) is the only one which draws attention to this fact.

The differences between the three presentations reflect differences in the authors' assessments of their intended audiences. None the less, all three versions share a common core. In each case the rainbow is explained in terms of a model in which a parallel beam of light is reflected by a set of refracting spheres. The behaviour of the light is then modelled in terms of ray theory. Some of this behaviour is summarized by the laws of refraction and reflection of light, which are brought into play in paragraphs (B) and (E). For this reason, it is tempting to give a Hempelian analysis of the explanation, and to say that the explanandum was deduced from a set of laws and initial conditions, none of which was superfluous to the derivation. But, this temptation (like many others) should be resisted. In the first place, as Cartwright has pointed out (1983: 43–7), the simple form of Snell's law used here can only be applied to refraction within an optically isotropic medium; the universality we require in a law is absent. If, however, we think of Snell's law simply as part of the specification of a model that represents the behaviour of light in a restricted set of circumstances, no such requirement exists.

This brings me to my second point. The so-called 'laws' employed here are all expressed in terms of the particular model we are using. The angles of incidence, refraction, and reflection that enter our calculations are angles between the drop's surface and rays of light, represented by geometrical lines. Since the terms of the laws refer to elements of a theoretical model, there is no interpretation of the laws which is independent of theory. A similar point is made by Ernest Nagel (1961: 82–3); from our perspective, the passage where he does so ends with a splendidly ironic twist:

Experiments on beams of light passing through a given medium to a denser one show that the index of refraction varies with the source of the beam. Thus, a beam issuing from the red end of the solar spectrum has a different index of refraction than has a beam issuing from the violet end. However, the experimental law based on such experiments is not formulated in unquestionably observational terms (e.g. in terms of the visible colours of light beams) but in terms of the relation between the index of refraction of a light ray and its wave frequency.

There is also a further assumption about light concealed in paragraph (*D*) of the explanation. Intensities of light are assumed to be monotonically additive: the more rays emerging in a given direction, the greater the intensity of the light emitted in that direction. On a Hempelian analysis, this assumption would have to figure in the explanation as a law of nature. Yet, the fact that light intensities do not, in general, behave in this way is now so well established that no one would grant this assumption any nomological force. In contrast, within Newton's model, the assumption is perfectly acceptable. He writes ([1730] 1952: 2): 'By the Rays of Light I understand its least Parts, and these as well Successive in the same Lines, as Contemporary in several Lines.' That light has such 'least parts' is, on our account, just part of the theoretical definitions of the model he is working with. A natural inference from this 'least parts' assumption is that the intensity of light on a given surface is proportional to the numbers of such rays striking it per unit time.

There are, of course, models with greater empirical adequacy than Newton's, notably the wave model developed in the nineteenth century by Young, Fresnel, and their successors. An explanation of the rainbow in terms of the wave theory, generally known as 'Airy's theory', was available from about 1840. This theory explained more and was in closer agreement with observation than Newton's.[16] Yet, as Boyer records (1987: 309), ten, even twenty

---

[16] See Boyer 1987: 308.

years after the theory appeared, workers in the field expressed their frustration that textbooks still explained the rainbow in Newtonian terms. But, in their complaints we should distinguish two demands, one legitimate, the other not. The legitimate demand was for a textbook that presented Airy's theory. The illegitimate demand was that the Newtonian explanation be 'banished from the teaching of science'. The demand is illegitimate because, as Noakes and the PSSC recognized a century later, Newton still gave us an understanding of the rainbow. In doing so he did not explain all the phenomena associated with rainbows; he said nothing, for example, about supernumerary bows, coloured bands sometimes observed within the inner edge of the primary bow and beyond the outer edge of the secondary.[17] Without inconsistency, Noakes can acknowledge the omission, even as he proffers the minimum deviation account as his explanation of the rainbow. His section on the rainbow ends with the paragraph:

A complete theory of the rainbow embracing both the true and the supernumerary bows has been worked out, and according to this the angular radius of the primary bow should be a little less than the value given by the minimum deviation explanation and that of the secondary a little greater. (Noakes 1953: 120)

Noakes is fully aware that the 'complete theory of the rainbow' has little in common with the minimum deviation explanation. Indeed, a large part of his book is concerned with the wave model of light. None the less, in writing this paragraph, he is not suggesting that students should simply forget what they have read. He regards the minimum deviation explanation, as I do, a a genuine explanation, albeit one couched in terms of a theory of limited usefulness. How good an explanation it is, and in what respects, remains to be seen.

7.4. UNDERSTANDING

In 1963, Mary Hesse put forward 'a view of theoretical explanation as metaphorical redescription'.[18] This proposal has never received the attention

---

[17] These 'supernumerary bows' are due to diffraction, i.e. a wave effect. They had been remarked on by Witelo as early as the thirteenth century, and a careful observation of them was made by Benjamin Langwith, Rector of Petworth, in 1722 (see Boyer 1987: 277–8). Langwith makes the astute comment, 'I begin now to surmise ... that the suppos'd exact Agreement between the Colours of the Rainbow and those of the Prism is the reason why [this phenomenon] has been so little observed'. [18] Hesse 1965: 111; page references are to Hesse 1980.

it deserves.[19] Hesse draws on the account of metaphor given by philosopher, Max Black, on which a metaphor involves a non-literal description of a primary system in terms of a secondary system.

Thus, when Prospero says of his brother Antonio, a primary system, 'he was | The ivy which had hid my princely trunk, | And suck'd my verdure our on't', Antonia and Prospero are seen in terms of a secondary system, ivy and tree trunk. On this account, a metaphor does not merely compare the two systems; it is not replaceable by a statement of the points of similarity between the two. Metaphors are enlightening because they present us with a new framework of ideas within which to locate the primary.

Thus, metaphor has a cognitive function. A similar function is served, Hesse argues, by the use of models in science. In nineteenth-century physics, for example, when the transmission of light was taken as the primary system (the *subject* of the model in my terminology), it was represented as a wave motion in a luminiferous ether. By redescribing it in this way the behaviour of light was made understandable—that is, it was explained. Similarly, in the mid-twentieth century a plasma was represented by a collection of electrons in a sea of positive charge, and thereby explained why 'electrons of a few hundred or a few thousand electron volts lost energy in jumps when scattered through a thin metal foil' (Feynman, Leighton, and Sands 1963–5: ii. 7.7)

The major problem for this account of explanation, and, I suspect, one reason why it has found few advocates, is that metaphors differ from scientific models in important respects. Metaphors, as Hesse points out, are intended to be understood. If a metaphor is to be successful, the secondary system must already be familiar to us, so that it arrives with a rich store of associations. No doubt some scientific models, like the fluid theories of heat and electricity in the nineteenth century, explain the unfamiliar in terms of the familiar. But, very often the reverse is true. To repeat a point that others have made before me, scientific explanation typically explains a familiar occurrence in terms of a markedly less familiar model.[20] We are familiar with the fact that a lump of iron usually stays in one piece instead of being 'crumbled out againe to his Atomies'.[21] Yet, only quantum mechanics can explain why it does so. And, the problem is very pronounced for someone who countenances the use of

---

[19] For example, it was not included in Joseph Pitt's anthology, *Theories of Explanations* (1988), nor is it mentioned in Salmon's (1989) overview of accounts of explanation.

[20] They are *like* metaphor, however, in that by their use the primary is de-familiarized—in the sense in which that word is used by Victor Shklovsky and by Bertolt Brecht.

[21] John Donne, 'An Anatomy of the World—The First Anniversary' (1611).

explanation of highly abstract models. The models, for example, of a New-tonian system, or of relativistic space-time, would normally be said to carry with it a rich store of associations; their abstractness serves to defamiliarize them. In 1963, when Hesse presented her proposal, the only models that philosophers of science discussed were analogue models. And, in the positivist climate of the time, such models were disreputable. To return to the problem raised by abstract models: in the absence of association evoked by a familiar secondary system, how does the abstract type of scientific model acquire the cognitive function that I ascribe to it? To answer this we need to look more carefully at the similarities and differences between models and metaphors.

They are similar in one key respect: each of them is a representation-as. The difference between them, crudely put, is that the secondary system of a metaphor is a pre-existing entity, whereas a scientific model of the type under discussion is a free creation. This characterization of the difference, however, requires immediate qualification. One the one hand, although a successful metaphor requires a familiar secondary system, not only are the ideas associated with it transferred to the primary system, but the complex of these associations is itself affected by the use of the metaphor. This is why Black calls his account of metaphor an 'interaction view'. An eighteenth-century painting by Joshua Reynolds may shed light on the view. The title of the painting is 'Mrs Siddons as the Muse of Tragedy'. But, just as Mrs Siddons represents the Muse of Tragedy, a wave model can represent light. For Joshua Reynolds, substitute Christiaan Huygens; for Mrs Siddons, the wave; for the Muse of Tragedy, light. And, reciprocally, just as the painting affects how we see Mrs Siddons, the adoption of the wave model of light affects not only the way scientists thought about light but the way they thought about waves.[22]

Notice that both these qualifications bring metaphors and modelling closer together, and to that extent undermine the original objection. For the sake of argument, however, I will take the extreme case of a mathematical model of a phenomenon, abstract enough for any associations that may have withered away. What, cognitively, takes their place? Both involve something I have not so far emphasized in this essay: the use of mathematics in physical theory.

When a scientific model is first presented to us, its applications are not immediately visible; we have to play with the model to put it to use. As an example of the process, consider how the wave theory of light is applied

---

[22] Mrs Siddons, the celebrated actress, was painted in consecutive years, first in 1784 by Joshua Reynolds, and then in 1785 by Thomas Gainsborough. The difference between the portraits is striking.

successively to the phenomena of interference, diffraction, and polarization. Some of these applications are worked through on our behalf, others are set as examples. In each case the hoped-for result is the same. We become the resources of the model. Whereas, in most cases, the significance of a metaphor takes only a brief time to sink, a model's resources are made available to us gradually as we learn how much can be deduced from its theoretical definition. As a succession of phenomena are shown to be successfully modelled, the models become familiar, and we become able to apply the theory ourselves; increasingly, we see the subject in terms of the model, and vice versa.

The individual episodes that increase our understanding are, typically, occasions when acts of explaining take place. (There are also occasions when acts of banging our heads against a wall take place.) We learn from a text, a teacher, or a fellow researcher, how theory can be applied to a specific topic. I take it to be a virtue of this account of explanation that it has the following corollary: the more we already know about Newton's theory of light, the more we will understand about the rainbow after reading his explanation of it. Another of its virtues is that it endorses our intuition that theoretical unification brings about an increase in understanding.[23]

Mathematical models, however, are not the only models that can provide explanations. In Essay 5, I drew attention to models which are themselves physical systems, and these models can be the most immediate sustentation of understanding. I will give four examples.

Example 1. In 1810, Thomas Young observed the behaviour of light that passed through two parallel slits before falling on a screen. His experiments can be reproduced using monochromatic light, and the result is that a series of bright and dark lines appear on the screen. If one assumes (as did Young) that light has a wave nature, this behaviour is due to the superposition of the two light waves emerging from the slits, the bright lines where the waves reinforce each other, the dark where they cancel each other out. The mathematical treatment of the behaviour is not arduous,[24] but neither is it trivial for someone who is innocent of trigonometric functions. Since about 1960, however, high school teachers of physics can make use of the ripple

---

[23] Friedman (1974: 195) takes the 'unifying effect of scientific theories' to be the 'crucial property' that gives rise to understanding.
[24] In Europe it is expected of a sixteen-year-old student who is specializing in physics.

tank, a flat-bottomed basin of water, together with a vibrating device with two prongs. The vibrator is lowered into the water, causing two sets of circular wave-fronts to advance across the surface of the water. Almost immediately a static fan-spaced pattern appears on the surface, regions of wave activity alternating with comparative calm. The spacing S between regions of wave activity is proportional to the distance from the vibrator; if the frequency of the vibrator is altered, the spacing S will be found to be inversely proportional to that frequency; if the first two-pronged vibrator is replaced with another, with a different distance between its prongs, the spacing S varies in inverse proportion to that distance. Each of the three relations obtained using water waves in the ripple tank corresponds to those predicted by the analytical treatment of the optical system. But, for many students understanding is more easily achieved by the material route than by the mathematical.

Example 2. In 1949, Richard Feynman published two papers which, by introducing the diagrams which bear his name, allowed quantum field theory to flourish in the 1950s and thereafter.[25] The originality of these diagrams was described by Freeman Dyson thus:

The usual way physics was done since the time of Newton was to begin by writing down some equations and then to work hard calculating solutions of the equations. … [Feynman] had a physical picture of the way things happen, and the picture gave him the solution directly with a minimum of calculation. It was a wonder that people who had spent their lives solving equations were baffled. Their minds were analytical; his was pictorial.

Feynman diagrams immediately found adherents; in 1967, eighteen years after their introduction, Richard Mattuck produced a textbook entirely devoted to these diagrams, showing, *inter alia*, how many applications had been found for them. At the end of the introductory chapter, Mattuck wrote:

Feynman diagrams have other appealing features beside their utility as a calculational tool. … [T]hey show directly the physical meaning of the perturbation term they represent. Another thing is that they reveal at a glance the structure of very complicated approximations by showing which sets of diagrams have been summed over. In this way, they have introduced a new language into physics. … And finally, one cannot be

---

[25] A Feynman diagram represents interactions between elementary particles and electromagnetic radiation within a two-dimensional Minkowski relativistic space-time. In these diagrams, while electrons are represented as travelling towards the future, positrons are represented as travelling towards the past.

immune to the Klee-like charm of the diagrams. Including in their ranks ... such characters as the 'necklace', the 'potato', and the 'tadpole', ... they constitute what might indeed be called 'perturbation theory in comic book form'.   (Mattuck 1967: 19)

For Dyson, Feynman 'had a physical picture of the way things happen'. For Mattuck, Feynman diagrams 'show *directly the physical meaning* of the perturbation term they represent' (my emphasis).

Example 3. This example involves material models of a very primitive nature; to wit, some pieces of stiff cardboard, each conforming to one of four shapes. Though finally explanatory, initially these pieces did not figure in an explanation. Instead, they were cut out within the context of discovery by James Watson in 1953, and played a key role in the quest by which, along with Francis Crick, he sought to establish the molecular structure of DNA. At that time, he and Crick were almost sure that components of DNA comprised a helical backbone and four bases: two purine bases (adenine and guanine) and two pyrimidine bases (thymine and cytosine). The question was how to integrate these bases with the helical backbone, possibly forming another helix (or two) in the process. To this end, Watson cut out cardboard scale models of these four bases, and on his desktop assembled pairs of cardboard 'bases', each pair held together by 'hydrogen bonds'. He quickly found that by putting together like with like he got nowhere. As he wrote later:

[I] began shifting the bases in and out of various pairing possibilities. Suddenly I became aware that an adenine-thymine pair held together by two hydrogen bonds was identical in shape to a guanine-cytosine pair held together by at least two hydrogen bonds. All the hydrogen bonds seemed to form naturally; no fudging was required to make the two types of base pairs identical in shape. ... I suspected that we now had the answer to the riddle of why the number of purine residues exactly equalled the number of pyrimidine residues. ... Furthermore, the hydrogen-bonding requirement meant that adenine would always pair with thymine, while guanine would only pair with cytosine. Chargaff's rules then suddenly stood out as a consequence of a double-helical structure for DNA.   (Watson and Crick 1968: 194–6)

Kinesthetic experience, rather than theoretical cogitation, may often lead to understanding.

Example 4. Prima facie, Newton's explanation of the rainbow, with its invocation of the laws of refraction and reflection, conforms to Hempel's deductive-nomological account of an explanation. But, there is an element of the explanation that Hempel would not have had room for, but without which the argument would be barely understandable. It is the diagram.

7.5. EVALUATION

In the introductory section of this essay, I summarized my account of theoretical explanation as follows:

We explain some feature X of the world by displaying a theoretical model M of part of the world, and demonstrating that there is a feature Y of the model that corresponds to X, and is not explicit in the definition of M.

This bald summary can now be amplified. The features of the world I am concerned with are general regularities. I have ignored problems that arise in connection with explanations of individual events, particularly the problem of explaining the occurrence of an event of low probability. These are the problems that statistical relevance models and causal accounts of explanation were designed to deal with. In so far as my discussion deals with probability at all, it does so by showing how statistical laws might be explained.

In section 7.2, I distinguished constitutive, analogue, and foundational models. These examples were chosen to show not only that there are different kinds of models, and that the theories associated with them can differ widely in domain, but, more importantly, that the distinction sometimes drawn between scientific models and foundational theory is better regarded as a distinction between kinds of models; common to all them is the fact that they should be thought of as representations. However, the boundaries between the categories I specified are not definite, nor is the list of categories exhaustive. Models of processes, of the formation of a rainbow, for instance, do not fit readily into any of them—though in the case of the rainbow it is clear that Newton's ray theory, or a wave theory of light, will function as a foundational theory.

The term 'demonstration', as it appears in the summary above, is used in its seventeenth-century sense, and is synonymous with our present-day 'mathematical proof'. The limiting case of a mathematical proof, however, is the repetition of one of the premises, and, to avoid explanation by stipulation, we disallow appeals to features of the model explicit in its theoretical definition. We do not explain the rectilinear propagation of light merely by adopting a ray theory. This is not to say, of course, that because a principle has acted on a legitimate foundational premiss within explanations, one should not look for an explanation for it.

My summary account of theoretical explanation spoke to its form. It said nothing about what made an explanation good or bad, nor did it raise the question whether, and in what circumstances, a proffered explanation might conform to the specification but nevertheless be rejected.

Two sets of considerations, one pragmatic, the other theoretical, govern our evaluation of an explanation. The pragmatic are the simpler to deal with. They are summed up by the obvious precept that an explanation should be pitched at a level appropriate to its intended audience. What to the most recent Nobel Laureate is a splendid explanation may be no explanation at all to me. It follows that the model appealed to in an explanation should be one with which the listener is familiar, or with which he could be reasonably expected to become familiar as the explanation proceeds. Thus, our best theory may not, in every situation, provide the best explanation.

But, by what standards are we to judge theories? Two criteria often canvassed by scientists and historians of science are, empirically, adequacy and explanatory power. Bohr's theory of the hydrogen atom, for instance, is held to have displayed considerable explanatory power, despite the fact that the only strikingly accurate predictions it made were the wavelengths of the lines of the hydrogen spectrum. It provided explanations for a dozen other thing, but for the most part got them right only within an order of magnitude.[26] It also seems plausible that, under certain circumstances, a theoretical innovation that accounted for something we already knew would be valued, even if it produced no new empirical knowledge. Imagine that a new generation of fundamental particles was produced, whose only novel contribution was to tell us why massless particles never carry electric charge. This fact we know already, and so the theory would not add to our empirical knowledge. Even so, would we not value it for the understanding it provided? My surmise is that, if the model the theory provided seemed a natural extension of one we already have on the books, or if, by postulating some new symmetry, let us say, it used a strategy with which we were already familiar, then we would. In enabling us to understand a significant fact, the theory would give an explanation of it.

None the less, the proposal that explanatory power should be considered an aim of physics, on a par with, but not identifiable with, empirical adequacy, has sometime met with resistance.[27] This resistance derives in part from a suspicion that an insistence on explanation is a covert, and sometimes

---

[26] See Hughes 1990b: 77–9.   [27] Pierre Duhem comes to mind (*PT*, 7–9).

not so covert, demand that our representations of natural processes should be governed by a set of a priori—even (perish the thought) metaphysical—commitments. But, the proposal I canvassed in the previous section, that theoretical explanation be viewed as metaphoric redescription, contains no metaphysical agenda. It suggests that theoretical explanations allow us to understand the world, not by showing its conformity to principles external to the theory, but by representing it in terms of the model the theory itself supplies. As we become aware of the resources of these representations, so we come to understand the phenomena they represent. Hence, as I argued earlier, the greater the variety of the contexts in which we see theory applied, the deeper this understanding becomes.

To help to sever disreputable metaphysical associations from legitimate explanations, a parallel can be drawn between linguistic and scientific understanding. Linguistic understanding is manifested in terms of competence: our ability to recognize whether a word is correctly applied, to paraphrase sentences in which the word appears, and to apply it ourselves in appropriate, but not wholly familiar, situations. Similarly, our understanding of theory is displayed by the ease with which we follow explanations and find analogues for the behaviour of the models to which they apply. We may also learn to apply the theory ourselves, albeit at the humble level of one who points out that, on the minimum deviation account of the rainbow, only the inner edge of the primary bow will be a pure colour.

We may accept the notion of explanatory power, and agree to measure it by the range and diversity of phenomena representable within the theory. We are still left, however, with the second question. Explanatory for whom? How are we to justify our disregard of the pragmatic dimensions of explanations? And, if the answer is that, in talking of explanatory power, we make an implicit reference to an ideal audience, the first problem merely makes way for another. We may grant that, for an ideal audience the wave theory of light has greater power than Newton's ray theory; but, given Noakes' intended audience, he was surely quite right to present his explanation of the rainbow in the terms of the latter.

One answer is this. Among the theories that deal with a certain set of phenomena, we distinguish, first of all, a *best* theory, and secondly, a class of *acceptable* theories. Explanatory power (with respect to an ideal audience) is one of the criteria by which we choose our best theories, the other being empirical adequacy. (I do not mean to suggest that the choice is always straightforward.) An acceptable theory is one that bears a relation, yet to be specified, to our best theory of a set of phenomena. When a number of competing explanations of a phenomenon are available, we should, on any specific occasion, choose the one we judge to be pitched at the level appropriate to the intended audience. Should there be more than one such explanation, we should choose the one provided by the better theory, again assessed in terms of empirical adequacy. Explanations in terms of unacceptable theories are to be rejected.[28]

To complete this account, I need to indicate what it is for a theory to be acceptable, but, before doing so I will give a general idea of what the notion involves. Models, like paintings, are neither true or false; instead, they offer more or less adequate representations of the world. To say that a theory is acceptable is to say that, within a restricted ambit, the representations it offers are adequate. The perspective from which we make this judgement is the one offered by our *best* theory. From this perspective, we may find the ontology of another theory extremely dubious, but nevertheless regard the models provided by that theory as perfectly adequate for certain purposes. The advent of general relativity, for example, may make us harbour doubts about the assumptions underlying Newton's theory of gravity. Yet, it would be lunatic to deny the adequacy of Newton's theory for virtually all the purposes that concern, for example, NASA. The fate of Newtonian mechanics should also remind us that, *sub specie aeternitatis*, our present best theory can only be regarded as acceptable, despite the fact that, for now, it is the one by which others are judged.

Unacceptable theories include the cluster of ether theories in the late nineteenth century and the Bohr–Kramers–Slater theory of radiation in the 1920s. In contrast, Newton's emission theory of light, though flawed, is in many contexts acceptable. It is acceptable from the perspective of the wave theory of light because a wave model can mimic the behaviour of a model in which light travels in straight lines that bend towards the normal on entering an optically denser medium. From our present viewpoint, the wave theory is in turn acceptable because quantum field theories can mimic the behaviour of a model in which transverse waves are propagated through the electromagnetic field. Notice that in each of the 'better' theories just

---

[28] Louise Antony first made me realize what now seems to me obvious: that not all explanations are acceptable.

given, what is being mimicked is behaviour that is typically appealed to in explanations provided by the 'acceptable' theory.

Acceptability, relative to a 'better' theory, does not imply reducibility, as that term was used, and so avoids the difficulties that beset that concept. There is certainly no requirement that the vocabulary of one theory be translatable into the vocabulary of the other.

8

The Discourse of Physics Exemplified by the Aharonov–Bohm Effect

> Discourse in general, and scientific discourse in particular, is so complex a reality that we not only can, but should, approach it at different levels and with different methods.
>
> Michel Foucault[1]

> Each speech genre in each area of speech communication has its own typical conception of the addressee, and this defines it as a genre.
>
> Mikhail Bakhtin[2]

PREAMBLE

The critic Ernest Jones has provided a Freudian analysis of Shakespeare's *Hamlet*, Georg Lukacs a Marxist reading of Scott's *Heart of Midlothian*, and Hans Reichenbach a logical reconstruction of Einstein's General Theory of Relativity. Each of these projects provided a valuable way of thinking about its object of study, but none of them exhausted what can be said about it. In this essay, I will employ M. M. Bakhtin's 'speech genres' to bring out aspects of a paper in the *Physical Review*, 'Significance of Electromagnetic Potentials in the Quantum Theory', co-authored in 1959 by Yakir Aharonov and David Bohm.

Bakhtin writes:

Language is realized in the form of individual concrete utterances (oral and written) by participants in the various areas of human activity. These utterances reflect the specific conditions and goals of each such area, not only through their content (thematic) and linguistic style, that is, the selection of the lexical, phraseological, and grammatical resources of the language, but above all through their compositional

---

[1] Foucault 1970: xv.   [2] Bakhtin 1986: 60.

structure. All three of these aspects—thematic content, style, and compositional structure—are inseparably linked to the *whole* of the utterance and are equally determined by the specific nature of the particular sphere of communication. Each separate utterance is individual, of course, but each sphere in which language is used develops its own *relatively stable type* of these utterances. These may be called *speech genres.* (Bakhtin 1986: 60; emphasis in the original)

Notice that speech genres are associated with spheres of human activity; in fact, '[E]ach sphere of human activity contains an entire repertoire of speech genres, that differentiate and grow as the particular sphere develops and becomes more complex' (ibid.). Furthermore, these spheres of human activity are also spheres of communication. As an example of a primitive and undeveloped sphere of activity we may take the parade ground, whose speech genres are limited to two: the genre of formulaic commands (in which, as Bakhtin points out, even the vocal inflections are prescribed by the genre) and the marginally more inventive genre of profane and scatological abuse.

In this essay I engage in two interlocking projects. I begin by looking at parts of a specific scientific utterance, approaching the text at various levels, as Foucault recommends. I refer to these as *the structural level, the theoretical level, the rhetorical level,* and *the level of discourse.* From this last level I move to the second project, an examination of the characteristics and function of a particular speech genre, which I call the *discourse of physics.* The utterance I start from is the by Yakir Aharonov and David Bohm. For economy I refer to it as the *A–B paper.* The speech genre of the next five sections of this chapter will be what the French call *explication de texte.*

8.1. THREE LEVELS OF ASSESSMENT

8.1.1. The A–B Paper: Preliminaries

I begin with the title: 'Significance of Electromagnetic Potentials in the Quantum Theory'. It tells us that the authors assume that their readers will know something about electromagnetic theory and something about quantum theory. It also suggests questions and comparisons: Is the significance of electromagnetic potentials in quantum theory going to be compared with the significance of some other potentials in the same theory, or with that of the same potentials in a different theory? The intended readers, those who habitually read the *Physical Review*, would assume the latter. They already know what the first paragraph of the paper tells them, that in classical electromagnetic theory the significance of potentials is moot. But, as they read on, they will find that Aharanov and Bohm also compare the significance of two different potentials in the same theory—namely, quantum mechanics.

The *Abstract* and the *Introduction* of the paper review three reasons for thinking that in classical theory potentials have no physical significance. (1) They are eliminable from the fundamental equations of the theory. (2) They play no causal role; the forces that the theory postulates are due to fields. (3) Classical electromagnetic theory is gauge invariant.

The last phrase needs some explication. In classical theory, potentials are very curious quantities. They have no absolute magnitudes. Take, for example, electric potential. We can set the zero of a scale of electric potential at any place we wish. Given a self-contained arrangement of electrically charged objects, all at different potentials, it makes no difference which of them is regarded as being at zero potential. In fact, none of them needs to be at zero potential. We can assign negative values to all of them, or positive for that matter, or we can label some of them negative and the others positive. What matter are the differences in their potentials. Provided these differences were the same in each case, all of them would be physically equivalent. Classical electromagnetic theory also invokes another potential, the magnetic potential **A**, usually called the *vector potential*. With regard to this there is even greater latitude in the assignments of values which can be regarded as physically equivalent to one another. This feature of vector potentials is referred to as *gauge invariance*.

Historically, there were good reasons for thinking of the vector potential **A** as just an artefact emerging from the vector calculus used in articulating the theory. In classical theory, the three characteristics of potentials that Aharonov and Bohm draw attention to—gauge invariance, their eliminability from the equations of the theory, and the fact that they play no causal role—all go hand in hand. Indeed, there are plausible arguments to suggest that each of them entails the others. But, in quantum theory this is not the case. While the equations of quantum mechanics are gauge invariant, potentials cannot be eliminated from them, as Aharonov and Bohm point out. In section 2 of their paper, they go on to argue that neither the electric nor the vector potential is devoid of causal power. With hindsight, we may say that they predict the *Aharonov–Bohm effect*.

8.1.2. The Structure of Section 2

Section 2 of the A–B paper is entitled, 'Possible Experiments Demonstrating the Role of Potentials in the Quantum Theory'. Its thematic structure is that of the thirty-two-bar popular song, the songs that became jazz standards, songs like Jerome Kern's 'Smoke Gets in Your Eyes', Rodgers and Hart's 'Bewitched, Bothered, and Bewildered', and Duke Ellington's 'Don't Get Around Much Any More'. All three share a common structure: musically they are divided into four eight-bar segments. The first two segments and the fourth are musically and thematically alike, but, in the third, variously known as *the bridge* or *the middle eight*, the melody modulates into a new key from which the composer extricates himself, more or less felicitously as the case may be, in time for the return of the original tune in bar twenty-five. This musical shift is often accompanied by a change in perspective. In 'Smoke Gets in Your Eyes', for example, the first two segments open with the lines, 'They asked me how I knew ...' and 'They said some day you'll find ...', respectively, whereas the third begins, 'But I chaffed them and I gaily laughed ...'. After this brief display of self-assertion, however, the final segment reverts to, 'Now laughing friends deride ...', thereby confirming the ontological dependence of the individual on the gaze of the Other, as we say in the trade.

In the same way, section 2 of the A–B paper is divided into just four segments. The first opens with the words, 'Suppose we have a charged particle in ...'. It describes a situation in which a charged particle is placed in a region where changes in potential take place but no fields exist. So does the second: 'Now consider a more complex experiment ...'. In each case, quantum mechanics predicts a change in the phase of the wave function that governs the particle's behaviour. In the third segment, the paper modulates into relativity theory: 'From relativistic considerations, it is easily seen that ...'; and then a natural segue takes us back to segment four, where the words, 'This corresponds to another experimental situation ...', introduce yet another situation in which quantum theory predicts that a change of potential will bring about a change of the wave function of a particle.

8.1.3. The Theoretical Level

I will go through the theoretical analysis in more detail, although, truth to tell, there is very little theory involved. I will sketch the theoretical moves in a bit more detail, although, again, there is very little theory involved. Segments

1 and 2 involve the use of Schrödinger's equation and not much else. The equation governs the evolution of the state of a quantum system with time:

$$i\hbar \partial \psi / \partial t = H\psi$$

Reading from right to left: ψ is the wave function representing the state and H is an operator acting on ψ to yield another wave function, which we denote by $H\psi$. In quantum theory, operators represent physical properties of a system; H represents its energy, and is known as the Hamiltonian of the system; $\partial \psi / \partial t$ is the rate of change of ψ with time; $\hbar$ is Planck's constant; and i is the square root of -1. Not only does this equation give the time evolution of the state function ψ with time, but it also acts as a constraint on the set of the system's states. Only those functions ψ that satisfy the equation can be states of the system. Thus, a change in the Hamiltonian will not only affect the evolution of the system's state, but will also change the set of allowable states for the system. Note also that the Hamiltonian H can contain terms for the potential energies of the system.

In segment 1 of this section, Aharonov and Bohm consider a charged particle in a Faraday cage. This is a region where we can change the electrical potential without introducing electric fields; the interior of a hollow conductor provides an example. If a varying voltage $\phi(t)$ is applied to the cage, an additional energy term $e\phi(t)$ must be added to the original Hamiltonian H_0 to produce the Hamiltonian $H_0 + e\phi(t)$, which Aharonov and Bohm write as $H_0 + V(t)$. The effect of this is that, for every state ψ_0 that was a solution for the old Hamiltonian H_0, there will be a new state $\psi_0 e^{-iS/\hbar}$ that is a solution for the new one. The expression $e^{-iS/\hbar}$ is a *phase factor*. The quantity S, usually referred to as an *action*, is a function of V: $S = \int V(t) \, dt$. Note that S is also a function of time.

The electrical potential $\phi(t)$ thus produces a change in the wave function, but, as Aharonov and Bohm remark, this change of phase will not produce any physical results.

In segment 2 this result is applied to a more complex arrangement (see Fig. 8.1). An electron beam is divided into two parts, and each part passes through a metal tube. Each tube effectively forms a temporary Faraday cage, and, as in segment 1, potentials applied to the two tubes will bring about changes in the wave functions associated with the two parts of the beam. Different potentials, or potentials that vary differently with time, will bring

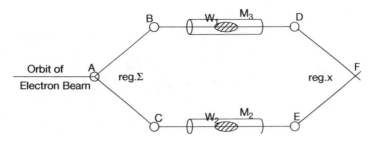

Fig. 8.1. Schematic experiment to demonstrate interference with time-dependent scalar potential. *A, B, C, D, E*: suitable devices to separate and divert beams. W_1, W_2: wave packets. M_2, M_3: cylindrical metal tubes. *F*: interference region.

about different changes in the wave functions ψ_1 and ψ_2 that represent the two parts of the beam, and the result will be an interference effect when the beams are recombined. If S_1 and S_2 are the two corresponding actions, then the interference effect will depend on the difference $(S_1-S_2)/\hbar$. The expressions for S_1 and S_2 are:

$$S_1 = e\int\phi_1\,dt \quad S_2 = e\int\phi_2\,dt$$

Leaving aside segment 3 for the moment, I will move to segment 4. Again, we have a divided beam, but this time the two parts pass on either side of a solenoid carrying an electric current (see Fig. 8.2). The magnetic field due to the current is confined within the solenoid, but it produces a vector potential **A** in the regions outside it through which the two parts of the beam travel. Aharonov and Bohm claim (at the top of p. 489) that, as in segment 2, there will be phase changes in the wave functions ψ_1 and ψ_2 associated with the two parts of the beam; further, that a difference in the two changes will produce an interference effect. As before, the two changes are associated with two actions, S_1 and S_2, and the interference effect will depend on $(S_1-S_2)/\hbar$.

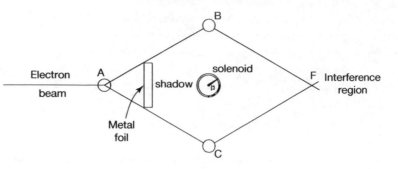

Fig. 8.2. Schematic experiment to demonstrate interference with time-independent vector potential.

Now however, this difference will again be given by an expression involving the vector potential **A**:

$$(S_1 - S_2)/\hbar = (e/\hbar c)\int \mathbf{A}.d\mathbf{x}$$

A quantity that was thought to be a mathematical artefact has acquired physical significance. The Ahanorov–Bohm effect has been predicted.

The difference of actions $(S_1 - S_2)/\hbar$ plays a key role both in segment 2 and in segment 4. Segment 2 provides an expression for this difference in terms of electric potentials, segment 4 an expression in terms of the vector potential. The licence to move from one to the other is supplied by the special theory of relativity in segment 3, the middle eight.

An aside on special relativity. As Minkowski pointed out, the special theory of relativity replaced independent theories of space and time with a single theory of space-time. Events are given four coordinates t, x_1, x_2, x_3, one of time and three of space. Space and time are seen as two sides of the same coin. The three-component vectors of Newtonian space and classical mechanics acquire an additional, temporal component to become 4-vectors. To the three components, p_1, p_2, p_3, of the momentum of a particle is added a fourth, its energy E, to form the energy-momentum 4-vector. The equations governing the transformations of the energy-momentum 4-vector $(E,\mathbf{p})$ under a change of coordinate system exactly match those governing the transformation of $(t,\mathbf{x})$ under that change.

Relativistically, the electrical potential ϕ of a point in space-time is the scalar, temporal component of the covariant 4-potential. Its spatial components are supplied by the (magnetic) vector potential $\mathbf{A}$. In the third segment of section 2 of their paper, Aharonov and Bohm abstract from the details of the thought experiment of segment 2 and invite the reader to think of the two paths followed by the wave-packets of that experiment as a single closed loop in space-time. They then equate the phase difference $(S_1-S_2)/\hbar$ of the two wave functions with the integral $(e/\hbar) \int \phi dt$ round that loop. The two terms in that integral are, respectively, the electrical potential ϕ at a point in space-time, and an infinitesimal time interval. As the authors point out, this integral can be regarded as a special case of its relativistic generalization: $(e/\hbar) \int [\phi dt - (\mathbf{A}/c).d\mathbf{x}]$ (c is just the conversion factor from standard units of length to light-seconds). Another special case is a closed loop in space-time where t remains constant. Associated with that loop, they argue, there should be associated a phase shift $\Delta S/h$ given by $(e/ch) \int \mathbf{A}.d\mathbf{x}$. And, with that remark, they return from the middle eight to the original key.

8.1.4. The Rhetorical Level

So much for the theoretical level of the argument. I now want to examine it at the rhetorical level, to look at the strategies of persuasion employed, and to see how the first two segments of section 2 prepare the ground for the fourth. As I have already noted, from a formal perspective these three segments are very much alike. In each case, an *action*, S, is defined for a system in an environment where a potential exists but no field is present. Schrödinger's equation is then used to show that when the potential is applied a change of phase of the system's wave function will occur, given by the expression $e^{-iS/\hbar}$. It is then claimed that, in the experimental arrangements described in segments two and four, the phase changes will produce observable effects. From another perspective, however, the segments are very different. To the original readers of the paper, the three conclusions would have differed markedly in their initial plausibility. In fact, the sequence in which they are presented is cunningly arranged to move from the unexceptionable to the unexpected.

8.2. THE LEVEL OF DISCOURSE

8.2.1. Shakespeare and the *Commedia dell'arte*

> Suppose within the girdle of these walls
> Are now confin'd two mighty monarchies,
> Whose high upreared and abutting fronts
> The perilous narrow ocean parts asunder[.]

So speaks the Chorus in the Prologue to Shakespeare's *Henry V*. Less poetically, Aharonov and Bohm ask us to 'suppose we have a charged particle in a Faraday cage', to 'consider a more complex experiment in which a single coherent electron beam is split into two parts, and each part is then allowed to enter a long cylindrical metal tube', and '[a]s another special case', to 'consider a path in space only' which 'corresponds to another experimental situation'.

Each of these invitations is a preamble to the representation of an action, the first to the performance of a play, the others to the printed narratives of a physics journal. Once the invitation is issued, the physicist, like the playwright, has the task of representing what is likely to happen—that is, what is capable of happening according to the rules of probability or necessity. But, the Prologue aside, the best theatrical analogy to the discourse of physics is provided not by Shakespeare's histories, but by the performances of the *commedia dell'arte*.

The *commedia dell'arte* was a form of improvizational theatre that flourished in Europe from the sixteenth to the eighteenth centuries. Stock characters improvised in stock situations. A typical piece might include two low-life characters, Brighella (a crafty schemer), and Arlecchino (simpleminded and half-starved), two bourgeois figures, the Dottore (pedantic and ineffectual) and Pantalone (lustful and avaricious), and, of course, a pair of young lovers, the *inamorato* and the *inamorata*, often called Isobella. All were immediately recognizable to the audience by the costumes they wore, and, in the case of the first four, by their half-masks.

The stock characters of the discourse of physics are the electron, the photon, the neutron, the pi-meson, and so on. They, too, find themselves in stock situations, imprisoned within a Faraday cage, for example, or in the vicinity of a current-carrying solenoid. Here, and in what follows, I use the phrase 'the discourse of physics' in a rather narrow sense. Used in a broader sense, the discourse of physics is constituted by all the utterances found in physics. It includes the observation reports that first indicated the existence of pulsars; it also includes the latest formulations of string theory. The utterances to which I shall apply the term occupy a discursive space intermediate between these two examples, a space where experimenter and theorist can communicate with each other. This is the arena in which the stock characters of physics play out their roles.

8.2.2. Galison and Trading Zones

Peter Galison (1997) describes spaces like this as 'trading zones'. He takes the term from anthropology: a trading zone is an area in which two cultures meet for purposes of barter and exchange. The coordination required for these activities may be achieved by using a pidgin, a simplified language that borrows elements from the languages of both cultures. Initially limited in expressive power, a pidgin may develop into an extended pidgin, and thence to a creole, a language rich and powerful enough to function as a first language for its speakers (Galison 1997: 48).

For Galison, contemporary physics involves interactions between a variety of cultures, some experimental, some theoretical, some concerned with instrumentation. He argues persuasively that Kuhn's picture of a shared paradigm, informing the assumptions, practices, and ontological commitments of theorist and experimenter alike, does not correspond to the way physics is carried on. On Galison's view, theory, experimentation, and instrumentation are quasi-autonomous domains, each with its own traditions and subcultures. The areas where these cultures interact with one another are the trading zones of physics. They may also be physical spaces, like the Radiation Laboratory at the Massachusetts Institute of Technology (MIT) that Galison describes (ibid. 817–27), but they need not be. As he remarks (ibid. 816), 'Trading between theorist and experimenter in the heyday of electron theories [the 1930s] was done by mail'.

Presumably, journals too can act as trading zones, provided they are not addressed just to the members of a specific subculture. The *Physical Review*, for instance, describes itself on its masthead as 'A journal of experimental and theoretical physics'. One typical volume in the 1950s included a report of measurements of the nuclear spins of isotopes of silver (Eubank et al.) and another of experiments to determine the optical properties of hexagonal ZnS crystals (Keller and Pettit); it also contained a paper setting out a classical theory of electronic and ionic inelastic collisions (Grysinski) and another describing the behaviour of a theoretical model, the charged ideal 2-dimensionsional Bose gas (May). Needless to say, I did not choose this volume of the *Physical Review* at random. It is the volume in which the Aharonov–Bohm paper appeared.

Though written by two theoretical physicists, the A–B paper is addressed to theorists and experimenters alike. In three out of the four segments of section 2, for example, the authors describe thought experiments involving, first, a single charged particle, and then a divided electron beam; the theoretical predictions they make are embedded in an experimental context. Sections 3 and 4 of the paper then deal, respectively, with experimental and theoretical issues. In section 3, the authors enquire whether the attainable separation of the parts of the beam is large enough to accommodate a solenoid or, failing that, a magnetic whisker, and, in section 4, they derive an exact solution of the Hamiltonian for an electron beam encountering a narrowly confined magnetic field. It is section 2, however, that prompts both of these investigations and explains how they are related. In this section, we may expect to find displayed the discourse of physics, in the narrow sense of the phrase.

Aharonov and Bohm present narratives; they tell us what would happen to an electron in certain situations. In doing so, they assume very little about their central character. I have already commented on how little theory is involved in section 2 of the paper. The authors assume that an electron can be treated as a quantum system, that a coherent beam of them can be described in terms of a wave function ψ, and that the evolution of ψ is governed by Schrödinger's equation. Notice that the question of whether or not the wave function ψ should be properly applied to a single electron (as in the first segment), or to a statistical ensemble of electrons (as in the second and fourth) is left to one side. This happens, not because the authors take this question lightly, but because foundational issues of this kind play no part in the discourse. Just as, for the audience of the *commedia dell' arte*, the costume and the half-mask of the Dottore bespoke pedantry and ineffectiveness, so too, for the reader of the *Physical Review*, the term 'electron' invokes certain properties. These include, not only the general property of behaving as a quantum system, but also other, more specific, properties. The electron has a certain charge, mass, and intrinsic spin, and it obeys the Pauli exclusion principle. As and when these properties are appealed to, they are simply taken as given. Again, their origin and deeper theoretical significance receive no more attention in this discourse than do the roots of Pantalone's lust and avarice in the productions of the *commedia dell'arte*.

Like that of Pantalone, the behaviour of the electron is determined by a combination of character and circumstance. Both individuals can be manipulated by someone, like Brighella, who is artful enough to do so. In the A–B paper, only two such manipulations are treated theoretically, and both of them involve the same property, the electronic charge e. The existence of this charge means that the Hamiltonian for the electron must be modified in standard ways when electromagnetic potentials are present. When an electrical potential ϕ is applied to a Faraday cage, the Hamiltonian H of an electron within it becomes $H + \phi e$; when an electron is moving through a region in which a vector potential $\mathbf{A}$ is present, a Hamiltonian $\mathbf{P}^2/2m$ (which would represent purely kinetic energy) becomes $[\mathbf{P} - (e/c)\mathbf{A}]^2/2m$. Other manipulations are taken for granted. The production of a coherent electron beam, its division into two parts, the recombination of these parts to produce interference, all of these are taken to be established laboratory procedures. Aharonov and Bohm give us the analogue of the *scenario* of the *commedia dell'arte*, the chronological plot summary that was tacked up backstage during a performance, listing the sequence of scenes and the places where the *lazzi* were to be inserted. (A *lazzo* is a comic routine, like that of Arlecchino catching a fly.) The precise manner in which these activities are to be carried out, be they experimental or theatrical, is left to the performers.

8.2.3. Theory and Experiment: Tonomura, Frank, and Bohr

The A–B paper is an utterance within the discourse of physics. Like any speech genre, the discourse of physics has its own concept of the addressee. Indeed, for Bakhtin (1986: 95), it is this that defines it as a genre. In this case, the collective addressee is the readership of the *Physical Review*. It includes both theoretical and experimental physicists, and the utterances within the genre connect the analysis of the one with the practices of the other. The language it employs is shared by both parties, but it is not a language of protocol sentences, nor does it lend itself to logical reconstruction as a partially interpreted formal language. Instead, it provides narratives involving entities like electrons. Often referred to as 'theoretical entities', within the discourse they are treated entirely realistically. Mathematical theory is kept to a minimum, experimental methods are underdescribed. Each party in the trading zone takes the other on trust. Each assumes that within the other's culture there are established standards of rigour that make for reliability. Their common language is a local language in Galison's sense (1997: 49); many features of each party's work are excluded, but its resources are nevertheless rich enough to allow them to coordinate their activities.

The terms of the language may have different connotations for the two parties. The theoretical assumptions that inform an experimenter's methods may not coincide with those of the theorist, and the experimenters or the theorists may even differ among themselves. Yet, these differences do not mean that each of them is talking about a different 'electron'. Compare the half-masks that were used in the *commedia dell'arte*. Each half-mask was ambiguous, so that from different perspectives they suggested different characteristics. None the less, in each case the characteristics were those of a single character. Aharonov and Bohm view an electron, or an electron beam, from a theoretical perspective, and ascribe to it a wave function governed by Schrödinger's equation. The perspective of the experimenter is noticeably different. Here is Akira Tonomura describing the Aharonov-Bohm effect in terms of the rotation of a wavefront:

An incident electron is deflected by the Lorentz force during its passing through a uniform magnetic field. Let's regard the electron as a wave. Since the wavefront is perpendicular to the electron trajectory the incident parallel electrons correspond to a plane wave, and the transmitted electrons to an inclined plane wave. In other words, the inclined wavefront is rotated round a magnetic line of force.   (Tonomura 1990: 23)

Two paragraphs later he redescribes what happens in terms of the vector potential:

This wavefront rotation may more easily be seen in terms of vector potentials $\mathbf{A}$. Since magnetic fields $\mathbf{B}$ are described as $\mathbf{B} = \mathrm{rot}\mathbf{A}$, the vector potential is circulating around the magnetic line. Since the vector potential shifts the incident wavefront, the wavefront level differs on both sides of the magnetic line [of force] but is the same along the magnetic line.   (ibid.)

Both parties, Aharonov and Bohm on the one hand and Tonomura on the other, use the term 'electron'. But, whereas for Aharonov and Bohm the background theory in terms of which the effect is described is Schrödinger's wave mechanics, for Tonomura it is the old (pre-1925) quantum theory, in which matter, like light, has both wave and particle properties. Despite their differences, neither party is in any danger of misunderstanding the other. Tonomura is, of course, well acquainted with wave mechanics, and Aharonov and Bohm can accept Tonomura's way of describing the effect as a perfectly intelligible *façon de parler*. Both parties recognize that they are discussing the same object—namely, the electron. They also share what Ian Hacking calls a 'common lore', rather than a common theory, about it, and it is this lore that informs their discourse.

Two aspects, at least, of this discourse would have offended Philipp Frank, a physicist closely affiliated with the Vienna Circle and with the logical empiricists in general. His monograph, *Foundations of Physics*, which appeared in volume 1 of *The International Encyclopedia of Unified Science* (edited by Otto Neurath, Rudolf Carnap, and Charles Morris), contains a section on quantum mechanics, in which he laments (1938: 481–2) 'the predilection of many authors for an "ontological" way of speaking', and 'the confusion ... produced [in discussions of that theory] by speaking of an object instead of the way in which words are used'. He also disapproves (ibid.) of the way in which 'the old language slips easily into the presentation of recent physics'. As we have seen, both these tendencies are displayed within the discourse. All parties speak unapologetically about electrons, and Tonomura uses 'the old language' of pre-1925 quantum theory to describe the Aharonov–Bohm effect. But, *pace* Frank, those are precisely the features of the discourse that allow it to function as an interlanguage between the culture of the theorist and that of the experimenter. By using 'an "ontological" way of speaking'—by talking about electrons rather than about the way the word 'electron' is used—the parties establish a common subject matter, to wit, electrons. By using a mode of description taken from early quantum theory, Tonomura can speak as an experimenter, in terms of cause and effect. In his first description of the Aharonov–Bohm effect, 'an incident electron is deflected by the Lorentz force' and 'the inclined wavefront is rotated round a magnetic line of force'; in the second, 'The vector potential shifts the incident wavefront'.

Simply put, these features of the discourse provide an answer to Bohr's problem. He writes:

Notwithstanding refinements of terminology due to accumulation of experimental evidence and developments of theoretical conceptions, all account of physical experience is, of course, ultimately based on common language, adapted to orientation in our surroundings and to tracing relationships between cause and effect. (Bohr 1963: 1)

Written late in his life, this is Bohr's most careful statement of a thesis to be found in many of his lectures and writings. He sees it as raising a problem of language. Bohr held classical theory to be an extension of our everyday discourse. Given the quiddities of quantum theory, he asks (ibid. 3) how may we establish 'the conditions for the unambiguous use of the concepts of classical physics in the analysis of atomic phenomena?' Instead, I suggest that the ambiguities raise a problem to which language provides the solution: How are the theorist and the experimenter to coordinate their activities? They do so by using a language in which alternative representations of the electron's behaviour, quantum-theoretical and classical, are both permissible.

8.2.4. Coda

I have described a particular form of scientific discourse, and have claimed that its features enable it to fulfil a specific function, that of acting as an *interlanguage* between two cultures of physics, the theoretical and the experimental. I will finish with two further suggestions: the first, that the general features of this discourse can be found in an overlapping group of speech genres (as Bakhtin would put it) within the physics community, and the second, that certain of its specific features limit it to a particular historical context.

The first thesis should not be surprising. As I have described it, the function of the discourse of physics is to enable theorist and experimenter in a given field to coordinate their activities. It does so by providing a common language in which the theoretician can express the physical import of his theoretical conclusions, and the experimenter can describe the results of her experiments. But, interest in the work of a given theorist, or pair of theorists, is not confined to a particular group of experimenters, nor is the audience for a certain set of experimental results confined to a particular group of theorists. The kind of discourse I have outlined enables both parties to address the physics community at large. Dealing as it does with the behaviour of stock characters in specified situations, it can fulfil a variety of functions. It can be descriptive, giving an account of what is happening, or what theory predicts will happen, in particular circumstances. It can be explanatory, answering questions about how or why a certain phenomenon occurs, or may be expected to occur. It can be used, as I have said, to show the physical significance of existing mathematical theory, but it may also be used to redescribe a physical situation it terms which make it amenable to theoretical treatment. Likewise, for the experimenter, it may not only be used to describe what is going on in existing experiments, but it may also be used to bring a predicted phenomenon within the ambit of standard experimental practices.

The features of the discourse that allow it to fulfil these functions are very general. They include its use of narrative, its implicit realism, and its reliance on lore rather than theory. Another feature, which I have not so far mentioned, is its use of metaphor to express that lore. Within these metaphors the stock characters of physics dance their allotted roles. Electrons are: 'boiled off the cathode'; 'the bosons have eaten the Higgs field'; 'the gluons condense into a taffy-like substance ... creating a string-like object with quarks at either end'; 'the muon spends most of its time in the nucleus'; 'the electrons gobble up two magnetic flux quanta per electron'; 'by means of time-determining electric "shutters" the beam is chopped into wave packets'. More poetically, they are: 'Gespensterfelder [ghost fields] guiding the photons'.

Described at this level of generality, this type of discourse appears, not as a pidgin to be used within a narrow trading zone, but as a lingua franca within the entire physics community. But, is there such a thing as 'the physics community'? Peter Galison's work suggests that no such community exists. In *Image and Logic*, he stresses the compartmentalization that characterizes modern physics. He sees physics as divided into three major cultures whose domains are theory, experiment, and instrumentation. Within the culture of experimental physics, he focuses on the subculture concerned with particle physics; within that, in turn, he distinguishes two sub-subcultures, one using image-producing devices and seeking individual 'golden events', the other using counters and accumulating statistical data. He emphasizes the very local nature of the pidgins that enable different subcultures, or sub-subcultures, to communicate with one another. In doing so, however, he largely neglects the fact that within physics there is a kind of discourse that is very widely shared. Nearly every physicist, but very few non-physicists, can read *Physics Today*. This fact provides the answer to my question: What is the physics community? It is the addressee of *Physics Today*.

So much for the general features of the discourse. My focus, however, was on the discourse found in a particular context, in discussions of the Aharonov–Bohm effect. A discourse as circumscribed as this may be expected to show features that are strongly dependent on historical circumstance. And so it proves. Recall Bohr's problem. According to Bohr (1963: 3), no matter how abstract and mathematically arcane our theories may become, 'the description of the experimental arrangement and the recording of observations must be given in plain language, suitably refined by the usual physical terminology'. In the discourse used by Tonomura and by Aharonov and Bohm, the solution to Bohr's problem was provided by the ambiguity of the electron's half-mask. The electron has been with us for a century; the lore surrounding it includes elements, not only of wave mechanics, but also of the early quantum theory, according to which complementary classical properties can be ascribed to it, and even of the fully classical account used by J. J. Thomson in the 1890s, when he measured the ratio of its charge to its mass. These classical elements link theory to experiment. For later arrivals on the stage of physics, no such classical or quasi-classical theory exists. Communication between theorist and experimenter must use other resources, and the resources available will differ from one historical context to another. We cannot legislate in advance what form future solutions to the problem will take. What they are in existing cases is a matter for empirical research.

Bibliography

Abrikosov, A. A. (1988). *Fundamentals of the Theory of Metals*. Trans. A. Beknazarov. Amsterdam: North Holland.

Achinstein, Peter (1983). *The Nature of Explanation*. New York; Oxford: Oxford University Press.

Aharonov, Yakir, and David Bohm (1959). 'Significance of Electromagnetic Potentials in the Quantum Theory'. *Physical Review*, 115: 485–91.

Ahlers, Guestner (1980). 'Critical Phenomena at Low Temperature'. *Reviews of Modern Physics*, 52: 489–503.

Albrecht, Andreas, David Coulson, Pedro Ferreira, and Joao Magueijo (1996). 'Causality, Randomness, and the Microwave Background'. *Physical Review Letters*, 76: 1413–16.

Amit, Daniel J. (1984). *Field Theory, The Renormalization Group, and Critical Phenomena*. 2nd edn. Singapore: World Scientific.

Anandan, Jeeva S. (ed.) (1990). *Quantum Coherence*. Proceedings of the International Conference on Fundamental Aspects of Quantum Theory. Singapore: World Scientific.

Anderson, Philip W. (1958), 'Random-Phase Approximation in the Theory of Superconductivity'. *Physical Review*, 112: 1900–16 (repr. Pines 1962: 424–40).

—— (1972). *Science*, 177: 393–6.

Andrews, Thomas (1869). 'On the Continuity of the Gaseous and Liquid States of Matter'. *Philosophical of the Royal Society*, 159: 57.

Aristotle (1975). *Posterior Analytics*. Ed. and trans. Jonathan Barnes. Oxford: Clarendon Press.

—— (1984). *Physics*. Trans. R. P. Hardie and R. K. Gaye. In Jonathan Barnes (ed.), *The Complete Works of Aristotle*. Princeton, NJ: Princeton University Press, 315–466.

Austin, J. L. (1965). *How to do Things with Words*. ed. by J. O. Urson. New York: Oxford University Press.

Ayer, A. J. (ed.) (1959). *Logical Positivism*. New York: Free Press.

Ayer, J. J. (1956/1969). *The Problem of Knowledge*. Harmondsworth Middlesex, England; Baltimore, Maryland; Ringwood, Australia: Penguin Books.

Baird, Davis (2004). *Thing Knowledge*. Berkeley, Calif.: University of California Press.

—— (1995). 'Meaning in a Material Medium.' In D. Hull, M. Forbes, and R. Burian, eds., *PSA 1994*, 2: 441–51.

Bakhtin, Mikhail M. (1986). *Speech Genres and other Late Essays*. Trans. Vern W. McGee. Ed. Caryl Emerson and Michael Holquist. Austin, Tex.: University of Texas Press.

Balbus, Steven A., and John F. Hawley (1998). 'Instability, Turbulence, and Enhanced Transport in Accretion Disks'. *Reviews of Modern Physics*, 70: 1–53.

Balzarini, D., and K. Ohra (1972). 'Co-existence Curve of Sulfur Hexafluoride'. *Physical Review Letters*, 29: 840–5.

Balzer, Wolfgang, C. Ulises Moulines, and Joseph D. Sneed (1987). *An Architectonic for Science*. Dordrecht: Reidel.

Bardeen J., N. L. Cooper, and J. R. Schrieffer (1957). 'Theory of Superconductivity'. *Physical Review*, 108: 5: 1175–204.

Bardeen, John, and David Pines (1955). 'Electron-phonon Interaction in Metals'. *Physical Review*, 99: 1140–50.

Beltrametti, Enrico, and Giovanni Cassinelli (1981). *The Logic of Quantum Mechanics*. Reading, Mass.: Addison-Wesley.

Blinder, K. (1984a). 'Monte Carlo Investigation of Phase Transitions and Critical Phenomena'. In Cyril Domb and M. S. Green (eds), *Phase Transitions and Critical Phenomena*. London: Academic Press (1984): v. 1–105.

—— (1984b). *Applications of the Monte Carlo Method in Statistical Physics*. Berlin: Springer-Verlag.

—— and D. Stauffer (1984). 'A Simple Introduction to Monte Carlo Simulation and Some Specialized Topics'. In Blinder 1984b: 1–36.

Bocchieri, P., A. Loinger, and G. Siragusa (1979). 'Non-existence of the Aharonov-Bohm Effect, II'. *Il Nuovo Cimento A*, 51: 1–17.

Bohm, David, and David Pines (1951). 'A Collective Description of Electron Interactions. I: Magnetic Interactions'. *Physical Review*, 82: 625–34.

—— —— (1952). 'A Collective Description of Electron Interactions. II: Collective *vs* Individual Particle Aspects of the Interactions'. *Physical Review*, 85: 338–53 (repr. Pines 1962: 170–80).

—— —— (1953). 'A Collective Description of Electron Interactions. III: Coulomb Interactions in a Degenerate Electron Gas'. *Physical Review*, 92: 609–25 (repr. Pines 1962: 153–69).

—— and Eugene P. Gross (1949a). 'Theory of Plasma Oscillations, A: Origin of Medium-like Behavior'. *Physical Review*, 75: 1851–64.

—— —— (1949b). 'Theory of Plasma Oscillations, B: Excitation and Damping of Oscillations'. *Physical Review*, 75: 1864–76.

Bohr, Niels (1913). 'On the Constitution of Atoms and Molecules'. *Philosophical Magazine*.

—— (1963). Essays 1958–1962. *Atomic Physics are Human Knowledge*. Intercourse Publishers. John Wiley & Sons: New York and London.

Boyer, Carl B. ([1959] 1987). *The Rainbow: From Myth to Mathematics*, Princeton, New Jersey: Princeton University Press.

Bragg, W. L., and E. J. Williams (1934). 'The Effect of Thermal Agitation on Atomic Arrangement in Alloy'. *Proceedings of the Royal Society*, A145: 699–730.

Bricker, Paul, and R. I. G. Hughes (eds) (1990). *Philosophical Perspectives on Newtonian Science*. Cambridge, Mass.: MIT Press.

Bridgman, P. W. ([1927] 1961). *The Logic of Modern Physics*. New York: Macmillan.

Brout, Robert (1957). 'Correlation Energy of a High-density Gas: Plasma Coordinates'. *Physical Review*, 108: 515–17 (repr. Pines 1962: 198–200).

Bruce, Alastair, and David Wallace (1989). 'Critical Point Phenomena: Universal Physics at Large Length Scales'. In Paul Davies (ed.), *The New Physics*. Cambridge: Cambridge University Press: 236–67.

Brush, Stephen G. (ed.) (1965). *Kinetic Theory*. 2 vols. Oxford: Pergamon.

——— (1967). 'History of the Lenz-Ising Model'. *Reviews of Modern Physics*, 30: 883–93.

Buchwald, Jed K. (1989). *The Rise of the Wave Theory of Light*. Chicago: University of Chicago Press.

Burke, Arthur W. (1970). *Essays on Cellular Automata*. Urbana, Ill.: University of Illinois Press.

Carnap, Rudolf (1922). *Der Raum: Ein Beitrag zur Wissenschaftslehre. Kant-Studien*, Ergänzungshefte, 56. Berlin: Reuther and Reichard.

——— ([1930–1] 1959). 'The Old and the New Logic'. Trans. I. Levi. In A. J. Ayer (ed.), *Logical Positivism*. New York: Free Press: 133–46. Orig. pub. in *Erkenntnis* 1 (1930–1).

——— (1937). *The Logical Syntax of Language*. London: Routledge and Kegan Paul.

——— (1939). *Foundations of Logic and Mathematics*. Chicago: Chicago University Press (repr. Otto Neurath, Rudolf Carnap, and Charles W. Morris (eds) (1955), *International Encyclopedia of Unified Science*, i. Chicago: University of Chicago: 139–213).

——— (1956). *Meaning and Necessity: A Study in Semantics and Modal Logic*. Chicago and London: The University of Chicago Press.

Cartwright, Nancy (1983). *How the Laws of Physics Lie*. Oxford: Clarendon Press.

——— (1999). 'Models and the Limits of Theory: Quantum Hamiltonians and the BCS Models of Superconductivity'. In Mary S. Morgan and Margaret Morrison Morgan (eds), *Models as Mediators: Perspectives on Natural and Social Science*. Cambridge: Cambridge University Press: 241–81.

——— Tofic Shomar, and Mauricio Suarez (1994). 'The Tool-box of Science'. In William E. Herfel, Wladlyslaw Krajewski, Ilkka Niiniluoto, and Ryszard Wójcicki (eds), *Theories and Models in Scientific Processes*. Poznan Studies in the Philosophy of the Sciences and the Humanities, 44. Rodopi: Amsterdam.

Cassels, J. M. (1970). *Basic Quantum Mechanics*. London: McGraw-Hill.

Cassirer, Ernst ([1921] 1953). [*Substance and Function, and*] *Einstein's Theory of Relativity*. Trans. W. Swabey and M. Swabey. London: Kegan Paul.

Churchland, Paul, and Clifford A. Hooker (eds) (1985). *Images of Science: Essays on Realism and Empiricism, with a Reply from Bas C. van Fraassen*. Chicago: University of Chicago Press.

Clausius, Rudolf ([1857] 1965). 'The Nature of the Motion that We Call Heat'. *Brush*, 1: 11–34.

Cohen, Robert S., Michael Horne, and John Stachel (eds) (1997). *Experimental Metaphysics. Quantum Mechanical Studies for Abner Shimony*. i. Boston: Kluwer Academic Publisher.

Cole, J. R., and S. Cole (1973). *Social Stratification in Science*. Chicago: University of Chicago Press.

Danto, Arthur (1986). *The Philosophical Disenfranchisement of Art*. New York: Columbia University Press.

Davies, Paul (ed.) (1989). *The New Physics*. Cambridge: Cambridge University Press.

Descartes, René ([1637] 2001). *Discourse on Method, Optics, Geometry, and Meteorology*. Trans. P. J. Olscamp. Indianapolis, Ind.: Hackett.

——([1644] 1983a). *Principles of Philosophy*. Trans. V. R. Miller and R. P. Miller. Dordrecht: Reidel.

——(1983b). *Œuvres de Descartes*. Ed. Charles Adam and Paul Tannery. 11 vols. Paris: Librairie Philosophique J. Vrin.

Dirac, P. A. M. (1935). *The Principles of Quantum Mechanics*. 2nd edn. Oxford: Clarendon Press.

Domb, Cyril (1960). 'On the Theory of Cooperative Phenomena'. *Advances in Physics: Philosophical Magazine Supplement*, 9: 149–6.

——(1996). *The Critical Point*. London: Taylor and Francis.

——and M. S. Green (eds) (1976). *Phase Transitions and Critical Phenomena*. v. London: Academic Press.

——and Martin F. Sykes (1961). 'Use of Series Expansions for the Ising Model Susceptibility and Excluded Volume Problems'. *Journal of Mathematical Physics*, 2: 63–7.

Duhem, Pierre ([1914] 1991). *The Aim and Structure of Physical Theory*. Trans. P. P. Wiener. 2nd edn. Princeton, NJ: Princeton University Press.

Eagleton, Terry (1984). *The Function of Criticism: From* The Spectator *to Post-Structuralism*. London: Verso.

Earman, John, and John D. Norton (1997). *The Cosmos of Science: Essays of Exploration*. Pittsburgh, Pa.: University of Pittsburgh Press/Universitätsverlag Konstanz.

Ehrenreich, H., and M. H. Cohen (1959). 'Self-consistent Field Approach to the Many-electron Problem'. *Physical Review*, 115: 786–90 (repr. Pines 1962: 255–9).

Einstein, A. ([1905] 1956). 'On the Movement of Small Particles Suspended in Stationary Liquid Demanded by the Molecular-Kinetic Theory of Heat'. In Einstein 1956: 1–18. Orig. pub. in *Annalen der Physik*, 17: 549–59.

—— ([1906a] 1956). 'A New Determination of Molecular Dimensions'. In Einstein 1956: 36–62. Orig. pub. in *Annalen der Physik*, 19: 289–306.

—— ([1906b] 1956). 'On the Theory of the Brownian Movement'. In Einstein 1956: 19–35. Orig. pub. in *Annalen der Physik*, 19: 371–81.

—— ([1908] 1956). 'The Elementary Theory of the Brownian Motion'. In Einstein 1956: 68–85. Orig. pub. in *Zeitschrift für Elektrochemie*, 14: 235–9.

—— ([1915] 1949). 'Explanation of the Perihelion of Mercury by the General Theory of Relativity'. Trans. R. I. G. Hughes. In *The Structure and Interpretation of Quantum Mechanics*. Cambridge, Mass.: Harvard University Press.

—— (1956). *Investigations on the Theory of Brownian Movement*. New York: Dover.

—— (1996). *The Collected Papers of Albert Einstein*. vi. *The Berlin Years. Writings, 1914–1917*. Ed. A. J. Kox, Martin J. Klein, and Robert Schulmann. Princeton, NJ: Princeton University Press.

Ewald, P. P., and H. Juretschke (1952). 'Atomic Theory of Surface Energy'. In Robert Gomer and Cyril Stanley (eds), *Structure and Properties of Solid Surfaces*. Chicago: University of Chicago Press: 82–119.

Feigl, Herbert (1956). 'Some Major Issues and Developments in the Philosophy of Science of Logical Empiricism'. In Herbert Feigl and Michael Scriven (eds) (1956): 3–37.

—— (1970). 'The "Orthodox" View of Theories: Remarks in Defense as well as Critique'. In Michael Radner and Stephen Winokur (eds), *Minnesota Studies in the Philosophy of Science*, iv. *Analyses of Theories and Methods of Physics and Psychology*. Minneapolis, Minn.: University of Minnesota Press, 3–16.

—— and Grover Maxwell (eds) (1961). *Current Issues in the Philosophy of Science*. New York: Holt, Rinehart and Winston.

—— and Michael Scriven (eds) (1956). *Minnesota Studies in the Philosophy of Science*, i. *The Foundations of Science and the Concepts of Psychology and Psychoanalysis*. Minneapolis, Minn.: University of Minnesota Press.

Fernandez, Roberto, Jürg Fröhlich, and Alan D. Sokal (1992). *Random Walks, Critical Phenomena, and Triviality in Quantum Field Theory*. Berlin: Springer-Verlag.

Feynman, Richard P., Robert B. Leighton, and Matthew Sands (1963–5). *The Feynman Lectures on Physics*, 3 vols. Reading, Mass.: Addison-Wesley.

Fine, Arthur, Mickey Forbes, and Linda Wessels (eds) (1991). *PSA 1990: Proceedings of the 1990 Biennial Meeting of the Philosophy of Science Association*. ii. East Lansing, Mich.: Philosophy of Science Association.

Fisher, Michael E. (1981). 'Simple Ising Models Still Thrive', *Physica*, 106A: 28–47. In Hahne 1983: 3448–71.

—— (1983). 'Scaling, Universality and Renormalization Group Theory'. 1–139. In Davis 1989: 3448–71.

Ford, Joseph (1989). 'What is Chaos that We Should be Mindful of It?'. In Paul Davies (ed.), *The New Physics*. Cambridge: Cambridge University Press: 248–71.

Foucault, Michel (1970). *The Order of Things: An Archaeology of the Human Sciences*. New York: Random House.

Frank, Philipp (1938). *Foundations of Physics*. In Otto Neurath, Rudolf Carnap, and Charles W. Morris (eds) (1955), *International Encyclopedia of Unified Science*, i (in 2 pts). Chicago: University of Chicago Press.

—— (1957). *Philosophy of Science: The Link Between Science and Philosophy*. Englewood Cliffs, NJ: Prentice Hall.

French, Peter A., Theodore E. Uehling Jr, and Howard K. Wettstein (eds) (1993). *Midwest Studies in Philosophy XVIII: Philosophy of Science*. Notre Dame, Ind.: University of Notre Dame Press.

Fried, Michael (1980). *Absorption and Theatricality: Painting and Beholder in the Age of Diderot*. Berkeley, Calif.: University of California Press.

Friedman, Michael (1983). *Foundations of Space–Time Theories: Relativistic Physics and Philosophy of Science*. Princeton, NJ: Princeton University Press.

—— (1988). 'Explanation and Scientific Understanding.' In *Theories of Explanations*. ed. Joseph C. Pitt. Oxford: Oxford University Press.

—— (1994). 'Geometry, Convention, and the Relativized A Priori'. In Wesley Salmon and Gereon Wolters (eds), *Logic, Language, and the Structure of Scientific Theories: Proceedings of the Carnap-Reichenbach Centennial, University of Konstanz, 21–24 May, 1991*. Pittsburgh, Pa. and Konstanz: University of Pittsburgh Press/Universitätsverlag Konstanz: 21–34.

Galilei, Galileo ([1638] 1974). *Discourses and Mathematical Demonstrations concerning Two New Sciences*, Trans. Stillman Drake. Madison, Wis.: University of Wisconsin Press.

Galison, Peter (1997). *Image and Logic: A Material Culture of Microphysics*. Chicago: University of Chicago Press.

Gardner, Martin (1983). *Wheels, Life, and other Mathematical Amusements*. New York: Freeman.

Gell-Mann, Murray (1957). 'Specific Heat of a Degenerate Electron Gas at High Density'. *Physical Review*, 106: 369–72 (repr. Pines 1962: 186–9).

—— and Keith A. Brueckner (1957). 'Correlation Energy of an Electron Gas at High Density'. *Physical Review*, 106: 364–8 (repr. Pines 1962: 181–5).

Georgi, Howard M. (1989). 'Effective Quantum Field Theories'. In Paul Davies (ed.), *The New Physics*. Cambridge: Cambridge University Press: 446–7.

Giere, Ronald M. (1985). 'Constructive Realism'. In Paul Churchland and Clifford A. Hooker (eds), *Images of Science: Essays on Realism and Empiricism, with a Reply from Bas C. van Fraassen*. Chicago: University of Chicago Press: 75–98.

—— (1996). 'From *Wissenschaftliche Philosophie* to Philosophy of Science'. In Ronald M. Giere and Alan W. Richardson (eds) (1996): 335–54.

—— and Alan W. Richardson (eds) (1996). *Origins of Logical Empiricism*, Minnesota Studies in the Philosophy of Science, 16. Minneapolis, Minn.: University of Minnesota Press.

Goldstone, J., and K. Gottfried (1959). 'Collective Excitations of Fermi Gases'. *Il Nuovo Cimento*, [X]13: 849–52 (repr. Pines 1962: 288–91).

Gomer, Robert, and Cyril Stanley (eds) (1952). *Structure and Properties of Solid Surfaces*. Chicago: University of Chicago Press.

Goodman, Nelson (1968). *Languages of Art*. Indianapolis, Ind.: Bobbs Merrill.

Griffiths, Robert B. (1957). 'Peierls' Proof of Spontaneous Magnetization of a Two-dimensional Ising Model'. *Physical Review*, A136: 437–9.

—— (1970). 'Dependence of Critical Indices upon a Parameter'. *Physical Review Letters*, 24: 1479–89.

Gross, Eugene (1987). 'Collective Variables in Elementary Quantum Mechanics'. In B. J. Hiley and F. David Peat (eds) (1987). *Quantum Implications: Essays in Honour of David Bohm*. London: Routledge and Kegan Paul: 46–65.

Grunbaum, A. and W. C. Salmon (eds) (1988). *The Limitations of Deductivism*. Berkeley, Calif.: University of California Press.

Guth, Alan and Steinhardt, Paul (1989). 'The Inflationary Universe.' In *The New Physics*. ed. Paul Davies. Cambridge; New York; Melbourne: Cambridge University Press.

Hacking, Ian (ed.) (1981). *Scientific Revolutions*. Oxford: Oxford University Press.

—— (1985). 'Do We See through a Microscope?'. In Paul Churchland and Clifford A. Hooker (eds), *Images of Science: Essays on Realism and Empiricism, with a Reply from Bas C. van Fraassen*. Chicago: University of Chicago Press: 132–52.

Hahne, E. J. W. (ed.) (1983). *Critical Phenomena*, Lecture Notes in Physics, 186. Berlin: Springer-Verlag.

Hartmann, Stephan (1996). 'The World as a Process'. In R. Hegselmann et al. (eds), *Modelling and Simulation in the Social Science from the Philosophy of Science Point of View*. Dordrecht: Kluwer: 77–100.

—— (1999). 'Models and Stories in Hadron Physics'. In Mary S. Morgan and Margaret Morrison (eds), *Models as Mediators: Perspectives on Natural and Social Science*. Cambridge: Cambridge University Press: 326–46.

Hegselmann, R. et al. (eds) (1996) *Modelling and Simulation in the Social Science from the Philosophy of Science Point of View*. Dordrecht: Kluwer.

Heisenberg, Werner (1925). 'Über quantentheoretische Umdeutung kinematischer und mechanischer Beziehungen'. *Zeitschrift für Physik*, 33: 879–93.

—— (1928). 'Theory of Ferromagnetism'. *Zeitschrift für Physik*, 49: 619–36.

Hempel, Carl G. (1965). *Aspects of Scientific Explanation and other Essays in the Philosophy of Science*. New York: Free Press.

Hempel, Carl G. (1966). *Philosophy of Natural Science*. Englewood Cliffs, NJ: Prentice-Hall.

—— (1970). 'On the "Standard Conception" of Scientific Theories'. In Michael Radner and Stephen Winokur (eds), *Minnesota Studies in the Philosophy of Science*, iv. *Analyses of Theories and Methods of Physics and Psychology*. Minneapolis, Minn.: University of Minnesota Press: 142–63.

—— (1988). 'Provisos: A Problem Concerning the Inferential Function of Scientific Theories'. In A. Grunbaum and W. C. Salmon (eds), *The Limitations of Deductivism*. Berkeley, Calif.: University of California Press.

—— and Paul Oppenheim (1948). 'Studies in the Logic of Explanation'. *Philosophy of Science*, 15: 567–79 (repr. Hempel 1965: 254–90).

Herring, Conyers (1952). 'Discussion Note on Ewald and Juretschke'. In Robert Gomer and Cyril Stanley (eds), *Structure and Properties of Solid Surfaces*. Chicago: University of Chicago Press: 117.

Hertz, Heinrich ([1894] 1956). *The Principles of Mechanics*. Intro. Robert Cohen; preface Hermann von Helmholtz. New York: Dover.

Hesse, Mary B. (1965, 111); page references are to Hesse (1980) edition. *Revolutions and Reconstructions in the Philosophy of Science*. Bloomington: Indiana University.

Hilbert, David ([1902] 1921). *Foundations of Geometry*. Chicago: Open Court.

Hiley, B. J. and F. David Peat (eds) (1987). *Quantum Implications: Essays in Honour of David Bohm*. London: Routledge and Kegan Paul.

Hoogland, A., J. Spaa, B. Selman, and A. Compagner (1983). 'A Special Purpose Processor for the Monte Carlo Simulation of Ising Spin System'. *Journal of Computational Physics*, 51: 250–60.

Howard, Don (1994). 'Einstein, Kant, and the Origins of Logical Empiricism'. Wesley Salmon and Gereon Wolters (eds), *Logic, Language, and the Structure of Scientific Theories: Proceedings of the Carnap-Reichenbach Centennial, University of Konstanz, 21–24 May, 1991*. Pittsburgh, Pa. and Konstanz: University of Pittsburgh Press/Universitätsverlag Konstanz: 45–105.

Hubbard, John (1957a). 'The Description of Collective Motion in Terms of Many-body Perturbation Theory'. *Proceedings of the Royal Society*, A240: 539–60 (repr. Pines 1962: 122–43).

—— (1957b). 'The Description of Collective Motion in Terms of Many-body Perturbation Theory. II: The Correlation Energy of a Free-Electron Gas'. *Proceedings of the Royal Society*, A243: 336–52 (repr. Pines 1962: 205–21).

Hugenholtz, N. M., and David Pines (1959). 'Ground State Energy and Excitation Spectrum of a System of Interacting Bosons'. *Physical Review*, 116: 489–506 (repr. Pines 1962: 332–49).

Hughes, R. I. G. (1989). *The Structure and Interpretation of Quantum Mechanics*. Cambridge, Mass.: Harvard University Press.

—— (1990a). 'Reason and Experiment in Newton's *Opticks*: Comments on Peter Achinstein'. In Paul Bricker and R. I. G. Hughes (eds), *Philosophical Perspectives on Newtonian Science*. Cambridge, Mass.: MIT Press: 75–184.

—— (1990b). 'The Bohr Atom, Models, and Realism'. *Philosophical Topics*, 18: 71–84.

—— (1993). 'Theoretical Explanation'. In Peter A. French, Theodore E. Uehling Jr, and Howard K. Wettstein (eds), *Midwest Studies in Philosophy*, 18: 132–53.

—— (1996). 'Semantic View of Theories'. *Encyclopedia of Applied Physics*, 17: 175–80.

—— (1997). 'Models, the Brownian Motion, and the Disunity of Physics'. In J. Earman and J. D. Norton (eds), *The Cosmos of Science: Essays of Exploration*. Pittsburgh, Pa. and Konstanz: University of Pittsburgh Press/Universitätsverlag Konstanz.

Hull, David, Mickey Forbes, and Richard M. Burian (eds) (1995). *PSA 1994: Proceedings of the 1994 Biennial Meeting of the Philosophy of Science Association*. ii. East Lansing, Mich.: Philosophy of Science Association.

Hume, David ([1776] 1947). *Dialogues Concerning Natural Religion*. Ed. Norman Kemp Smith. Indianapolis, Ind.: Bobbs Merrill.

Humphreys, Paul (1991). 'Computer Simulation'. In Arthur Fine, Mickey Forbes, and Linda Wessels (eds), *PSA 1990: Proceedings of the 1990 Biennial Meeting of the Philosophy of Science Association*. ii. East Lansing, Mich.: Philosophy of Science Association: 497–506.

Husserl, Edmund ([1954] 1970). *The Crisis of European Sciences and Transcendental Phenomenology: An Introduction to Phenomenological Philosophy*. Trans. David Carr. Evanston, Ill.: Northwestern University Press.

Ising, Ernst (1925). 'A Contribution to the Theory of Ferromagnetism'. *Zeitschrift für Physik*, 31: 253–8.

Jammer, Max (1966). *The Conceptual Development of Quantum Mechanics*. New York: McGraw-Hill.

Jones, G. A. (1987). *The Properties of Nuclei*. 2nd edn. Oxford: Clarendon Press.

Jungnickel, Christa, and Russell McCormmach (1986). *Intellectual Mastery of Nature: Theoretical Physics from Ohm to Einstein*. 2 vols. Chicago: University of Chicago Press.

Kaku, Michio (1993). *Quantum Field Theory: a Modern Introduction*. New York: Oxford University Press.

Kant, Immanuel ([1781] 1998). *Critique of Pure Reason*. Trans. and ed. P. Guyer and A. W. Wood. Cambridge: Cambridge University Press.

Kermode, Frank (1983). *The Classic: Literary Images of Permanence and Change*. Cambridge, Mass.: Harvard University Press.

Koestler, Arthur ([1959] 1968). *The Sleepwalkers: A History of Man's Changing Vision of the Universe*. Harmondsworth: Penguin.

Kubo, Ryogo, and Takeo Nagamiya (1969). *Solid State Physics*. Ed. Robert S. Knox. New York: McGraw-Hill.

Kuhn, Thomas S. (1957). *The Copernican Revolution*. Cambridge, Mass.: Harvard University Press.

Lakatos, Imre ([1970] 1978). *Philosophical Papers*. i. *The Methodology of Scientific Research Programmes*. Ed. John Worrall and Gregory Currie. Cambridge: Cambridge University Press.

Landau, L. D. ([1937] 1965). 'On the Theory of Phase Transitions'. In Landau (1965): 193–216.

—— (1965). *Collected Papers*. Ed. D. ter Haar. Dordrecht: Kluwer.

—— (1981). *Science and Hypothesis*. Dordrecht: Reidel.

——, and Lifshitz, E. M. (1958). *Statistical Physics*. Trans. from the Russian by E. Peierls and R. T. Peirls. London: Pergamon Press; Reading, Mass.: Addison-Wesley Publishing Company.

Landman, Uzi et al. (1990). 'Atomistic Mechanisms and Dynamics of Adhesion, Nanoindentation, and Fracture'. *Science*, 248: 454–61.

Leavis, F. R., and Q. D. Leavis (1969). *Lectures in America*. London: Chatto and Windus.

Leavis, Q. D. (1969). 'A Fresh Approach to *Wuthering Heights*'. In F. R. Leavis and Q. D. Leavis, *Lectures in America*. London: Chatto and Windus: 83–152.

Lee, E. W. (1963). *Magnetism*. Harmondsworth: Penguin.

Lewis, D. (1973). *Counterfactuals*. Cambridge, Mass.: Harvard University Press.

Longair, Malcolm (1989). 'The New Astrophysics'. In Paul Davies (ed.), *The New Physics*. Cambridge: Cambridge University Press: 94–208.

Machamer, Peter, Lindley Darden, and Carl Craven (2000). 'Thinking about Mechanisms'. *Philosophy of Science*, 67: 1–27.

Mattuck, Richard D. (1967). *A Guide Feynman Diagram in the Many-Body a Problem*. London; New York: McGraw-Hill.

Messiah, Albert (1956). *Quantum Mechanics*. Trans. G. M. Tenner. 2 vols. New York: John Wiley.

Misner, Charles W., Kip S. Thorne, and John Archibald Wheeler (1970). *Gravitation*. San Francisco, Calif.: W. H. Freeman.

Morgan, Mary S., and Margaret Morrison (eds) (1999). *Models as Mediators: Perspectives on Natural and Social Science*. Cambridge: Cambridge University Press.

Morgenbesser, Sidney (ed.) (1967). *Philosophy of Physics Today*. New York: Basic Books.

Morrison, Margaret (1992). 'A Study in Theoretical Unification: The Case of Maxwell's Electromagnetic Theory'. *Studies in the History and Philosophy of Science*, 23: 103–45.

——(1999). 'Models as Autonomous Agents.' *Models as Mediators: Perspectives on Natural and Social Science*. (p. 53–60). Cambridge; New York; Melbourne; Cambridge University Press.

Mott, Neville (1956a). *Atomic Structure and the Strength of Metals*. London: Pergamon Press.

——(1956b) 'Theoretical Chemistry of Metals'. *Nature*, 178: 1205–7.

Motz, Lloyd (1964) *Astronomy: A to Z*. trans., by Arthur Beer, Grosset & Dunlap, Inc., *Publishers*. New York.

Moulines, C. Ulises (1998). 'Structuralism vs. Operationalism'. In Gerhard Preyer, Georg Peter, and Alexander Ulfig (eds), *After the Received View: Developments in the Theory of Science. Proto-Sociology*, 12: 40–58.

Nagel, Ernest (1961). *The Structure of Science: Problems in the Logic of Scientific Explanation*. New York: Harcourt, Brace & World.

——Patrick Suppes, and Alfred Tarski (eds) (1962). *Logic, Methodology, and Philosophy of Science: Proceedings of the 1960 International Congress*. Stanford, Calif.: Stanford University Press.

Neurath, Otto ([1938] 1955). 'Unified Science as Encyclopedic Integration'. In Otto Neurath, Rudolf Carnap, and Charles W. Morris (eds) (1955), *International Encyclopedia of Unified Science*, i (in 2 pts). Chicago: University of Chicago Press: 1–27.

Newton, Isaac ([1686] 1934). *Philosophiae Naturalis Principia Mathematica*. Ed. F. Cajori. 3rd edn. Berkeley, Calif.: University of California Press.

——([1704] 1952). *Opticks: or A Treatise of the Reflections, Refractions, Inflections, and Colours of Light*, 4th edn. Foreword by Albert Einstein; introduction by Edmund Whittaker; preface by I. Bernard Cohen; and analytical table of contents prepared by Duane H. D. Roller. New York: Dover.

Niemeijer, T., and J. M. J. van Leeuwen (1974). 'Wilson Theory for 2-Dimensional Spin Systems'. *Physica*, 71: 17–40.

Noakes, G. R. (1957). *New Intermediate Physics*. London: Macmillan and Co., Ltd.

Nozières, Philippe, and David Pines (1958a). 'Correlation Energy of a Free Electron Gas'. *Physical Review*, 111: 442–54 (repr. Pines 1962: 222–34).

——— ——(1958b). 'Electron Interaction in Solids: Collective Approach to the Dielective Constant'. *The Physical Review*, 109: 762–77.

Oberdan, Thomas (1994). 'Comment: Einstein's Disenchantment'. In Wesley Salmon and Gereon Wolters (eds), *Logic, Language, and the Structure of Scientific Theories: Proceedings of the Carnap-Reichenbach Centennial, University of Konstanz, 21–24 May, 1991*. Pittsburgh, Pa. and Konstanz: University of Pittsburgh Press/Universitätsverlag Konstanz: 107–18.

Ohm, Georg Simon (1827). *Die Galvanische Kette, mathematische bearbeitet*. Berlin: Riemann.

Pais, Abraham (1982). *'Subtle is the Lord ...': The Science and the Life of Albert Einstein*. Oxford: Oxford University Press.

Pearson, Robert B., John L. Richardson, and Doug Toussaint (1983). 'A Fast Processor for Monte-Carlo Simulation'. *Journal of Computational Science*, 51: 241–9.

Peierls, Rudolf (1936). 'Ising's Model of Ferromagnetism'. *Proceedings of the Cambridge Philosophical Society*, 232: 477–81.

Perrin, Jean Baptiste (1913). *Les Atomes*. Paris: Alcan.

Petzoldt, Joseph (1921). *Die Stellung der Relativitätstheorie in der geistigen Entwicklung der Menschheit*. Dresden: Sibyllen-Verlag.

Pfeuty, Pierre, and Gerard Toulouse (1977). *Introduction to the Renormalization Group and to Critical Phenomena*. Trans. G. Barton. New York: John Wiley.

Pines, David (1953). 'A Collective Description of Electron Interactions. IV: Electron Interactions in Metals'. *Physical Review*, 92: 626–36 (repr. Pines 1962: 170–80).

—— (1955). 'Electron Interactions in Metals'. In Frederick Seitz and David Turnbull (eds), *Solid State Physics: Advances in Research and Applications*, i. New York: Academic Press: 367–450.

—— (1956). 'Collective Energy Losses in Solids'. *Reviews of Modern Physics*, 28: 184–98.

—— (1962). *The Many-Body Problem*. New York: Benjamin.

—— (1987). 'The Collective Description of Particle Interactions: from Plasmas to the Helium Liquids'. In B. J. Hiley and F. David Peat (eds), *Quantum Implications: Essays in Honour of David Bohm*. London: Routledge and Kegan Paul: 66–84.

Pitt, Joseph (ed.) (1988). *Theories of Explanations*. New York: Oxford University Press.

Platzman, P. M., and P. A. Wolff (1973). *Waves and Interactions in Solid State Physics*. New York: Academic Press.

Poincaré, Henri ([1902] 1952). *Science and Hypothesis*. Trans. W. S. Greensreet. New York: Dover.

—— (1908). *Science et méthode*. Paris: Flammarion.

Popper, Karl (1959). *The Logic of Scientific Discovery*. Trans. J. Freed and L. Freed. London: Hutchinson.

Preyer, Gerhard, Georg Peter, and Alexander Ulfig (eds) (1998). *After the Received View: Developments in the Theory of Science*. Proto-Sociology, 12.

Proust, Joëlle (1994). 'Comment [on Friedman (1994)]'. In Wesley Salmon and Gereon Wolters (eds), *Logic, Language, and the Structure of Scientific Theories: Proceedings of the Carnap-Reichenbach Centennial, University of Konstanz, 21–24 May, 1991*. Pittsburgh, Pa. and Konstanz: University of Pittsburgh Press/Universitätsverlag Konstanz: 35–43.

Putnam, Hilary (1962). 'What Theories are Not'. In Ernest Nagel, Patrick Suppes, and Alfred Tarski (eds), *Logic, Methodology, and Philosophy of Science: Proceedings of the 1960 International Congress*. Stanford, Calif.: Stanford University Press: 240–51.

—— ([1974] 1981). 'The "Corroboration" of Theories'. In Ian Hacking (ed.), *Scientific Revolutions*. Oxford: Oxford University Press: 60–79. Orig pub. in Paul A. Schilpp (ed.) (1974), *The Philosophy of Karl Popper*. 2 vols. Library of Living Philosophers. La Salle, Ill.: Open Court: i. 221–40.

Radner, Michael, and Stephen Winokur (eds) (1970). *Minnesota Studies in the Philosophy of Science*, iv. *Analyses of Theories and Methods of Physics and Psychology*. Minneapolis, Minn.: University of Minnesota Press.

Raimes, Stanley (1957). 'The Theory of Plasma Oscillations in Metals'. *Reports on Progress in Physics*, 20: 1–37.

—— (1961). *The Wave Mechanics of Electrons in Metals*. Amsterdam: North Holland.

Reichenbach, Hans ([1920] 1965). *The Theory of Relativity and A Priori Knowledge*. Ed. and trans. M. Reichenbach. Berkeley, Calif.: University of California Press.

—— ([1924] 1969). *Axiomatization of the Theory of Relativity*. Ed. and trans. M. Reichenbach. Berkeley, Calif.: University of California Press.

—— ([1928] 1957). *The Philosophy of Space and Time*. Trans. M Reichenbach and J. Freund. New York: Dover.

—— ([1951] 1964). *The Rise of Scientific Philosophy*. Berkeley, Calif.: University of California Press.

Reitz, John R. (1955). 'Methods of the One-Electron Theory of Solids'. In Frederick Seitz and David Turnbull (eds), *Solid State Physics: Advances in Research and Applications*, i. New York: Academic Press: 2–95.

Rohrlich, Fritz (1991). 'Computer Simulation in the Physical Sciences'. In Arthur Fine, Mickey Forbes, and Linda Wessels (eds), *PSA 1990: Proceedings of the 1990 Biennial Meeting of the Philosophy of Science Association*. ii. East Lansing, Mich.: Philosophy of Science Association: 507–18.

Roukes, Michael (2001). 'Plenty of Room, Indeed'. *Scientific American*, Sept.: 48–57 <http://209.85.229.132/search?q=cache:a0jNFP5qPsgJ:nano.caltech .edu/papers/SciAm-Sep01.pdf+the+unique+physics+of+the+nanoscale ,+Roukes&cd=2&hl=en&ct=clnk&gl=uk>.

Roy, S. M. (1980). 'Condition for Non-Existence of Aharonov-Bohm Effect'. *Physical Review Letters*, 44: 111–14.

Ruelle, David (1991). *Chance and Chaos*. Princeton, NJ: Princeton University Press.

Ruthemann, Von Gerhard (1948). 'Diskrete Energieverluste mittelschneller Elec-tronen beim Durchgang dünne Folien'. *Annalen der Physik*, 2: 113–34.

Ryckman, T. A. (1996). 'Einstein *Agonistes*: Weyl and Reichenbach on Geometry and the General Theory of Relativity'. In Ronald M. Giere and Alan W. Richardson (eds), *Origins of Logical Empiricism*, Minnesota Studies in the Philosophy of Science, 16. Minneapolis, Minn.: University of Minnesota Press: 165–209.

Ryle, Gilbert ([1949] 1963). *The Concept of Mind*. Harmondsworth: Penguin.

Sabra, A. I. (1981). *Theories of Light: From Descartes to Newton*. Cambridge: Cambridge University Press.

Sachs, Mendel (1963). *Solid State Theory*. New York: McGraw-Hill.

Salmon, Wesley (1984). *Scientific Explanation and the Causal Structure of the World*. Princeton, NJ: Princeton University Press.

——(1989). *Four Decades of Scientific Explanation*. Minneapolis, Minn.: University of Minnesota Press.

—— and Gereon Wolters (eds) (1994). *Logic, Language, and the Structure of Scientific Theories: Proceedings of the Carnap-Reichenbach Centennial, University of Konstanz, 21–24 May, 1991*. Pittsburgh, Pa. and Konstanz: University of Pittsburgh Press/Universitätsverlag Konstanz.

Schilpp, Paul A. (ed) (1974). *The Philosophy of Karl Popper*. 2 vols. Library of Living Philosophers. La Salle, Ill.: Open Court.

Schlick, Moritz (1915). 'Die philosophische Bedeutung des Relativitätsprinzips'. *Zeitschrift für Philosophie und philosophische Kritik*, 159: 129–75.

——([1917] 1920). *Space and Time in Contemporary Physics*. Trans. H. L. Brose. Oxford: Oxford University Press.

——([1930–1] 1959). 'The Turning Point in Philosophy'. Trans. D. Rynin. In A. J. Ayer (ed.), *Logical Positivism*. New York: Free Press: 53–9. Orig. pub. in *Erkenntnis*, 1 (1930–1).

Schrödinger, Erwin (1926). 'Quantisierung als Eigenwertproblem'. *Annalen der Physik*, 79: 361–76; 80: 437–90; 81: 109–39.

Schweber, Sylvan S. (1997). 'The Metaphysics of Science at the End of a Heroic Age'. In Robert S. Cohen, Michael Horne, and John Stachel (eds), *Experimental Metaphysics. Quantum Mechanical Studies for Abner Shimony*. i. Boston: Kluwer Academic Publisher: 171–98.

Schwinger, Julian (1948). 'Quantum Electrodynamics I: A Covariant Formulation'. *Physical Review*, 74: 1439–61.

——(1949a). 'Quantum Electrodynamics II: Vacuum Polarization and Self-energy'. *Physical Review*, 75: 657.

Schwinger, Julian (1949b). 'Quantum Electrodynamics. III: The Electromagnetic Properties of the Electron; Radiation Corrections to Scattering'. *Physical Review*, 76: 790–817.

Scriven, Michael (1961). 'The Key Property of Physical Laws: Inaccuracy'. In Herbert Feigl and Grover Maxwell (eds) (1961), *Current Issues in the Philosophy of Science*. New York: Holt, Rinehart and Winston: 91–101.

Seitz, Frederick (1960). *Modern Theory of Solids*. New York: McGraw-Hill.

—— and David Turnbull (eds) (1955). *Solid State Physics: Advances in Research and Applications*, i. New York: Academic Press.

Shakura, Nikolai, and Rashid Sunyaev (1973). 'Black Holes in Binary Systems: Observational Appearance'. *Astronomy and Astrophysics*, 24: 337–55.

Sklar, Lawrence (1992). *Philosophy of Physics*. Boulder, Col.: Westview Press.

——(1993). *Physics and Chance: Philosophical Issues in the Foundations of Statistical Mechanics*. Cambridge: Cambridge University Press.

Smith, C. M. H. (1966). *A Textbook of Nuclear Physics*. London: Pergamon.

Sneed, Joseph D. (1979). *The Logical Structure of Mathematical Physics*. 2nd edn. Dordrecht: Reidel.

Sommerfeld, Arnold (1928). 'Electron Theory on the Basis of the Fermi Statistics'. *Zeitschrift für Physik*, 47: 1–32.

Stalnaker, R. (1968). 'A Theory of Conditionals'. In N. Rescher (ed.), *Studies in Logical Theory*. American Philosophical Quarterly Monograph Series, 2: 98–122.

Stanley, H. Eugene (1971). *Introduction to Phase Transitions and Critical Phenomena*. Oxford: Clarendon Press.

Steele, Martin C., and Bayram Vural (1969). *Wave Interactions in Solid State Plasmas*. New York: McGraw-Hill.

Stegmüller, Wolfgang (1976). *The Structure and Dynamics of Theories*. New York: Springer-Verlag.

——(1979). *The Structuralist View of Theories*. New York: Springer-Verlag.

Suarez, Mauricio (1999). 'The Role of Models in the Application of Scientific Theories: Epistemological Implications'. In Mary S. Morgan and Margaret Morrison (eds), *Models as Mediators: Perspectives on Natural and Social Science*. Cambridge: Cambridge University Press: 168–96.

Suppe, Patrick (ed.) (1957). *Introduction to Logic*. Princeton, NJ: van Nostrand.

——(1960). 'A Comparison of the Meaning and Uses of Models in Mathematics and the Empirical Sciences'. *Synthese*, 12: 287–301 (repr. Suppes 1969: 10–23).

—— (1967). 'What is a Scientific Theory?'. In Sidney Morgenbesser (ed.), *Philosophy of Physics Today*. New York: Basic Books: 55–67.

——(1969). *Studies in the Methodology and Foundations of Science. Selected Papers from 1951 to 1969*. Dordrecht-Holland: Reidel.

——(1977a). *The Structure of Scientific Theories*. 2nd edn. Urbana, Ill.: University of Illinois Press.

——(1977b). 'The Search for Philosophical Understanding of Scientific Theories'. In Frederick Suppe (ed.) (1977a): 3–241.

——(1989). *The Semantic Conception of Theories and Scientific Realism*. 2nd edn. Urbana, Ill.: University of Illinois Press.

Tarski, Alfred (1965). *Introduction to Logic and to the Methodology of the Deductive Sciences*. 3rd edn. New York: Oxford University Press.

Thouless, David J. (1989). 'Condensed Matter Physics in less than Three Dimensions'. In Paul Davies (ed.), *The New Physics*. Cambridge: Cambridge University Press: 20–35.

Toffoli, Tommaso (1984). 'Cellular Automata as an Alternative to (rather than an Approximation of) Differential Equations in Modelling Physics'. *Physica*, 10D: 117–27.

Tomonaga, Sin-Itiro (1946). 'On a Relativistically Invariant Formulation of the Quantum Theory of Wave Fields'. *Progress in Theoretical Physics*, 1: 27.

——(1948). 'On Infinite Field Reactions in Quantum Field Theory'. *Physical Review*, 74: 224–5.

Tonks, Lewi, and Irving Langmuir (1929). *Physical Review*, 33: 195–210.

Tonomura, Akira (1990). 'Experimental Confirmation of the Aharonov-Bohm Effect by Electron Holography'. In Jeeva S. Anandan (ed.), *Quantum Coherence*. Proceedings of the International Conference on Fundamental Aspects of Quantum Theory. Singapore: World Scientific: 18–33.

Toulmin, Stephen ([1953] 1960). *The Philosophy of Science: An Introduction*. New York: Harper.

Trumpler, Maria (1997). 'Verification and Variation: Pattern of Experimentation in Investigations of Galvanism in Germany, 1790–1800'. *Philosophy of Science*, (suppl.): (64), pp. 1–10.

van Fraassen, Bas C. (1980). *The Scientific Image*. Oxford: Clarendon Press.

——(1985). 'Empiricism in Philosophy of Science'. In Paul Churchland and Clifford A. Hooker (eds), *Images of Science: Essays on Realism and Empiricism, with a Reply from Bas C. van Fraassen*. Chicago: University of Chicago Press: 245–308.

——(1989). *Laws and Symmetries*. Oxford: Clarendon Press.

——(1991). *Quantum Mechanics: An Empiricist View*. Oxford: Clarendon Press.

von Neumann, John ([1932] 1955). *Mathematical Foundations of Quantum Mechanics*. Trans. into English by Robert T. Beyer. Princeton: Princeton University Press.

Vichniac, Gérard Y. (1984). 'Simulating Physics with Cellular Automata'. *Physica*, 10D: 96–116.

Watson, James D., and F. H. C. Crick (1953). 'Molecular Structure of Nucleic Acids: A Structure for Deoxyribose Nucleic Acid'. *Nature*, 171: 737–8.

——(1968). *The Double Helix*. New York: Atheneum.

Wegener, Alfred (1912). 'Die Entstehung der Kontinente'. *Petermanns Geographische Mitteilung*, 58: 185–95, 253–6, 305–9.

Wehr, M. Russell, and James A. Richards Jr (1960). *Physics of the Atom*. Reading, Mass.: Addison-Wesley.

Weinberg, Steven (1972). *Gravitation and Cosmology: Principles and Applications of the General Theory of Relativity*. New York: John Wiley.

——(1994). *Dreams of a Final Theory*. New York: Random House.

Wentzel, G. (1949). *Quantum Theory of Wave Fields*. New York: Interscience.

——(1957). 'Diamagnetism of a Dense Electron Gas'. *Physical Review*, 108: 1593–6 (repr. Pines 1962: 201–4).

—— ([1953] 1974). *Philosophical Investigations*. Trans. by G. E. M. Anscombe and R. Rhees. Third edition of English and German. Basil Blackwell: Oxford.

Will, Clifford (1989). 'The Renaissance of General Relativity'. In Paul Davies (ed.), *The New Physics*. Cambridge: Cambridge University Press: 7–33.

Wigner, Eugene P. (1959). *Group Theory and its Application to the Quantum Mechanics of Atomic Spectra*. Trans. from German by J. J. Griffin. New York; San Francisco; London: Academic Press.

Wilson, C. (1968). 'Kepler's Derivation of the Elliptical Path'. *Isis*, 59: 5–25.

Wilson, Kenneth G. (1975). 'The Renormalization Group: Critical Phenomena and the Kondo Problem'. *Reviews of Modern Physics*, 47: 773–840.

Wittgenstein, Ludwig ([1921] 1974). *Tractatus Logico-Philosophicus*. Trans. D. F. Pears and B. F. McGuinness. London: Routledge.

Wolfram, Stephen (1984). Preface to *Physica*, 10D (spec. issue on cellular automata): vii–xiii.

Wolters, Gereon (1994). 'Scientific Philosophy: The Other Side'. In Wesley Salmon and Gereon Wolters (eds), *Logic, Language, and the Structure of Scientific Theories: Proceedings of the Carnap-Reichenbach Centennial, University of Konstanz, 21–24 May, 1991*. Pittsburgh, Pa. and Konstanz: University of Pittsburgh Press/Universitätsverlag Konstanz: 3–19.

Ziman, J. M. (1960). *Electrons and Phonons: The Theory of Transport Phenomena in Solids*. Oxford: Clarendon Press.

—— (1964). *Principles of the Theory of Solids*. Cambridge: Cambridge University Press.

—— (1978). *Reliable Knowledge: An Exploration of the Grounds for Belief in Science*. Cambridge: Cambridge University Press.

General Index